VERTICAL REEFS

Number Twenty-seven
GULF COAST BOOKS
Sponsored by Texas A&M University–
Corpus Christi

John W. Tunnell Jr., General Editor

A list of titles in this series is available at the end of the book.

VERTICAL REEFS

MARY KATHERINE WICKSTEN

Texas A&M
University Press
College Station

Life on
Oil and Gas
Platforms
in the
Gulf of Mexico

Manufactured in China by Everbest Printing Co.
through FCI Print Group
This paper meets the requirements of ANSI/NISO Z39.48–1992 (Permanence of Paper).
Binding materials have been chosen for durability.

LIBRARY OF CONGRESS CATALOGING-IN-PUBLICATION DATA

Wicksten, Mary K., author.
Vertical reefs : life on oil and gas platforms in the Gulf of Mexico /
Mary Katherine Wicksten.—First edition.
pages cm—(Gulf Coast books ; number twenty-seven)
Includes bibliographical references and index.
ISBN 978-1-62349-311-0 (flexbound : alk. paper)—
ISBN 1-62349-311-0 (flexbound : alk. paper)—
ISBN 978-1-62349-312-7 (e-book)
1. Drilling platforms—Environmental aspects—Mexico, Gulf of.
2. Marine ecology—Mexico, Gulf of. 3. Artificial reefs—Mexico, Gulf of.
4. Continental shelf—Mexico, Gulf of. I. Title. II. Series: Gulf Coast books ; no. 27.
QH545.D75W53 2015
577.7'272—dc23
2015004797

Unless otherwise noted all photographs by Mary K. Wicksten.

CONTENTS

PREFACE

On a clear night at sea, the bright golden lights of oil platforms sparkle in the Gulf of Mexico. People on the beach and most of the public call them "oil rigs." Some view them negatively as unsightly sources of pollution, others see them as valuable producers of much-needed fuel.

The thousands of oil platforms that sit off the coasts of the northern Gulf of Mexico are conspicuous three-dimensional features on what is mostly a flat, muddy sea floor. The plants and animals that live on or near them are, for the most part, different from those found beneath them or on the nearby shore. For resident or migratory underwater species, these platforms play an important role in their ecosystems. They also may play a lesser-known role in the migratory patterns of the birds and even insects that fly over them.

After years of experience diving on the oil platforms of the Gulf of Mexico, photographer Dick Zingula and I decided to write a book that explained the structure and function of oil platforms, and how the marine ecosystems of these structures compare with those of natural reefs. Emphasizing the areas most likely to be visited by anglers and recreational divers, from the surface to 130 feet (40 meters), we cover the common or easily seen species found on or near platforms, going from those situated closer to shore out to those in what is called the blue water. We hope that the book will be a helpful reference for the teacher, beachgoer, angler, diver or other member of the interested public. The area of coverage, the western Gulf, follows the Bureau of Ocean Energy Management (BOEM) designation of the area between the Texas-Louisiana border and the US-Mexican border.

The descriptions in this book concentrate on the platforms of the western Gulf of Mexico and the life found on them. Readers in more eastern or southern parts of the Gulf of Mexico or other parts of the world may see similar platforms, but it is likely that the fauna will be different. Common names used here are those established by the American Fisheries Society and American Ornithologists' Union. See the section on further information for published references. Many of the observations on natural history are my own, derived firsthand from diving on platforms in the High Island area of the Gulf and Buccaneer Field and shipwrecks and natural reefs off Texas. Most of the photographs were taken on platforms off Texas. Where a good quality photograph was not available, I have used a photograph of the same species from elsewhere in the Gulf of Mexico or Caribbean waters. Photographs of specimens show material from the Biodiversity Research and Teaching Collections (formerly the Texas Cooperative Wildlife Collection), Texas A&M University; and the Harte Research Institute for Gulf of Mexico Studies, Texas A&M University-Corpus Christi. Depths are given in feet and meters.

ACKNOWLEDGMENTS

Any work of this length benefits from the help and suggestions of colleagues. Dick Zingula, whose lovely photographs grace this book, has been an advocate for diving in the Gulf of Mexico for many years. His experience both in diving and the petroleum industry were valuable in preparation of this book. I would like to thank Gregg Gitschlag, National Marine Fisheries, Galveston, for his help in providing information on removal of platforms. Gregory Boland of the Bureau of Ocean Energy Management, Department of the Interior, provided valuable and hard-to-get references on ecological studies on the platforms. Jennifer Pallanich, US editor of *Offshore Engineer*, provided constructive criticism of an earlier draft of the manuscript. Benny Gallaway, LGL Inc. of Bryan, Texas, provided literature and helpful criticism. I also thank Page Williams and Dana Larsen of Houston, Texas, for their comments and Anne Rudloe of Gulf Specimen Marine Laboratories, Panacea, Florida, for an additional review. The captains and crews of the marine vessels *Spree* and *Fling* provided safe and enjoyable opportunities for diving on the platforms. Diving on HI A389-A was conducted, in part, during the "Down Under, Out Yonder" program of the Flower Garden Banks National Marine Sanctuary, G. P. Schmahl, director. Archie Ammons, T. J. Boyle, Dan Campbell, Linda Pequegnat, and Angela Witmer, of Texas A&M University; Noe Barrera and Fabio Moretzsohn, of the Harte Institute for Gulf of Mexico Studies; and photographers Quenton Dokken, Carol Cox, Cindy Abgarian, Esat Atikkan, Bill Samaras, Jesse Cancelmo, Terry Sohl, and Simon Pierce, granted the use of their photographs. The work profited from the many helpful suggestions of anonymous reviewers. I especially thank Lloyd Hetrick of British Petroleum for insisting, "It's a platform, not a rig!"

VERTICAL REEFS

THE WHY AND WHERE OF DRILLING FOR OIL AND GAS

THE FLOOR of the Gulf of Mexico consists of sand and mud that have settled out of the water column and been compressed over time. The Gulf of Mexico first opened in the Jurassic period, over 140 million years ago. For millions of years, it was connected with a vast inland sea that gradually retreated. Sea levels fluctuated, at times dropping to almost 600 feet (185 meters) lower than modern level with shorelines well offshore from their present position, and at other times rising high enough to reach far inland from the modern coast. Fossils of marine fishes, crabs, and even cuttlefishes that are fifty million years old can be found along riverbanks in Brazos County, Texas. While huge deposits of salt from the ancient seas accumulated under the more modern Gulf floor, silt, debris, and vegetation washed downstream from the Mississippi River and the Brazos, Sabine, Colorado, and Guadalupe rivers of Texas as well as other, smaller river systems. The weight of the sediment created pressure on the buried organic materials and, along with decomposition, produced heat. The buried material released gasses and gradually changed into petroleum compounds. The resulting collection of oil and gas eventually filtered into porous rock and usually remained trapped there, caught beneath salt deposits or other rock that formed a cap on top of it.

Today, some natural oil seeps in the Gulf of Mexico produce slicks or globules that float to the surface, but most of the deposits remain deeply buried. Some oil is hidden under extensive salt deposits. Instead of drilling, geologists today can locate potentially usable deposits by seismic profiling—emitting vibrations, small explosions, or bursts of air from a ship and detecting the layering of sediments made by their reflection.

Judging from the gentle slope and relatively flat appearance of the shoreline of the western Gulf, it might be assumed that the sea floor is also relatively flat. But the continental shelf (which extends from the shoreline into the Gulf where waters are less than 600 feet, or about 185 meters deep), varies in width from over 100 miles off Galveston to less than 50 miles off South

Padre Island. Elevated areas, either banks or reefs, occur here but do not reach the surface. These elevated areas can be a clue to the existence of an oil or gas deposit because they may be evidence that pressure from the oil or gas has forced up buried layers of sediment. Some banks or reefs, such as Seven and One-Half Fathom Reef, lie within 25 miles of the shore. A series of banks at the edge of the shelf runs from the West Flower Garden Bank eastward to Sonnier Bank off Louisiana. The depth of these banks varies, from those easily accessible to scuba divers (such as the East Flower Garden Bank, which reaches to within 65 feet [20 meters] of the surface) to well below normal diving limits (such as Diaphus Bank, at roughly 240 feet or 73 meters).

The topography of the Gulf of Mexico continues to vary in the deeper waters past the continental shelf. In the western Gulf, there are elevations and also depressions (called basins) as well as the Alaminos Canyon and other submarine canyons, which can indicate regions of ancient salt and petroleum deposits. At the base of the continental slope down to more than 10,840 feet (3,300 meters) deep in the southwestern Gulf of Mexico lies the Sigsbee Abyssal Plain.

Recent studies and mapping by oil companies and oceanographic expeditions have shown that gas deposits are common throughout the Gulf. Like petroleum, natural gas (mostly methane) forms from decomposing organic matter. Natural gas may be associated with petroleum deposits or may exist by itself. Even at only 90 feet (30 meters) at the West Flower Garden Banks, gas bubbles emerge from cracks on the sea floor. At depths of over 900 feet (273 meters), gas hydrates are found. Similar to ice, these deposits incorporate methane in a "cage" of water molecules. The temperature at which these compounds form depends on the water pressure and the gas composition, but generally is just above the freezing point of seawater (about 41 °F, or 2 °C). Tubeworms, blind squat lobsters, deepwater mussels, and other odd animals lie near these deposits. Understanding the strange communities of bacteria and animals associated with the gas seeps, and determining if they are unique to particular areas or endangered by human activities, is the subject of much research.

People have long utilized both oil and natural gas. Early peoples of the Middle East, as well as the original inhabitants of California, used tar to seal their boats, waterproof their baskets, and attach shell beads to bowls and ornaments. Tar was used to waterproof European boats and ropes and to discourage wood-boring organisms from damaging the hulls of sailing ships. The use of oil and gas for fuel began to increase in the late 1800s when the availability of whale oil, commonly used to fuel lamps, began to decline due to over-hunting of these animals. Exploration for petroleum deposits began in earnest. Drilling began on land, where surface seeps were visible. In Texas, major drilling began at the famous Spindletop Field, south of Beaumont, in 1892. Early shallow drilling could not reach the main deposits, but deeper drilling in 1901 produced "gushers." The resulting oil boom attracted "wildcatters" and other colorful characters as Houston, Beaumont, and other Texas cities prospered. Although independent oil and gas operators remain, over time many operations consolidated into the familiar large oil corporations of today. As the older fields began to play out, the exploitation of underwater petroleum deposits became a focus of many oil companies.

The first offshore well was drilled off the coast of Summerland, California, in 1896. Early exploration for the most part was confined to piers, with pipes extending to the shoreline. In 1938, the first offshore well in the Gulf of Mexico was developed in 13–14 feet (about 4 meters) of water off Calcasieu Parish, Louisiana. But as the demand for oil and gas grew, exploration moved into deeper waters. In November 1947, the first well completely out of sight of land in the Gulf was drilled 12 miles south of Terrebone Parish, Louisiana. New technologies allowed drilling from ships and the industry began to establish sturdy platforms out at sea. Experiences with oil spills at this stage contributed to the rapid development of techniques to seal pipes and prevent leakage by the industry.

Growing worldwide demand for oil and gas provided an economic impetus for exploration and production in ever-deeper areas. In 1991, Shell Oil Company installed the world's largest standing structure, the Bullwinkle Platform, which is 1,736 feet (526 meters) tall and located in 1,350 feet (409 meters) of water. By 1997, production in the Gulf continued down to depths past 5,000 feet (1,525 meters). In 2003, an exploratory well was drilled near the Alaminos Canyon at 10,011 feet (3,034 meters) deep. Today, the Perdido Development, located over 200 miles from the Texas coast at a water depth of more than 1.2 miles (6,336 feet or 1,931 meters), is the world's deepest offshore drilling and production facility.

Currently, about 21 percent of all the natural gas and 30 percent of oil produced in the United States comes from wells in the Gulf of Mexico. Petroleum constitutes a major fuel source, but it also is used to produce lubricants and wax, and is a component in the making of plastics and other chemicals. Natural gas is used for fuel and to generate power and heat, and is also an essential component of fertilizers. The website of the US Bureau of Ocean Energy Management (BOEM) Gulf of Mexico Region provides up-to-date information and yearly reports on oil and gas exploration and production in the Gulf of Mexico.

The Leasing Process

After an oil company has determined that potential oil or gas deposits exist in an area, what happens next? Petrochemical companies cannot drill in shipping lanes, or in areas that have been used for military operations. The states of Texas and Louisiana have control of the seabed from the shoreline to a distance of one to three leagues (about 3–10 miles). In Louisiana, the Petroleum Lands Division of the Office of Mineral Resources manages leasing; in Texas, the General Land Office, as part of the Permanent School Fund, controls the bidding. In both states, an interested company nominates a defined tract of sea floor for production. The company provides a sealed bid, including a detailed plan of operations and location. In Louisiana, the State Mineral and Energy Board considers bids at regular meetings after review by the Office of Mineral Resources. In Texas, the School Land Board holds sealed bid sales every four months. Bids for leases are awarded upon terms "appearing most advantageous to the state" (according to the Louisiana Office of Mineral Resources "Leasing Manual"). Additional requirements for ensuring safety measures, means of reporting and mitigating oil spills, and sealing of wellheads must be fulfilled before leases are awarded. Each state receives royalties from oil or gas produced on state lands.

Since 1953, oil and gas resources located beyond that distance have been under the control of the BOEM, a part of the US Department of the Interior. The area under the control of the BOEM (federal waters) is called the outer continental shelf (OCS). A company bids for the right to explore and extract petroleum products from a particular geographic area, called a block.

Bidding today is a five-year process. BOEM calls for nominations of areas of interest from potential bidders. A notice of intent to prepare an

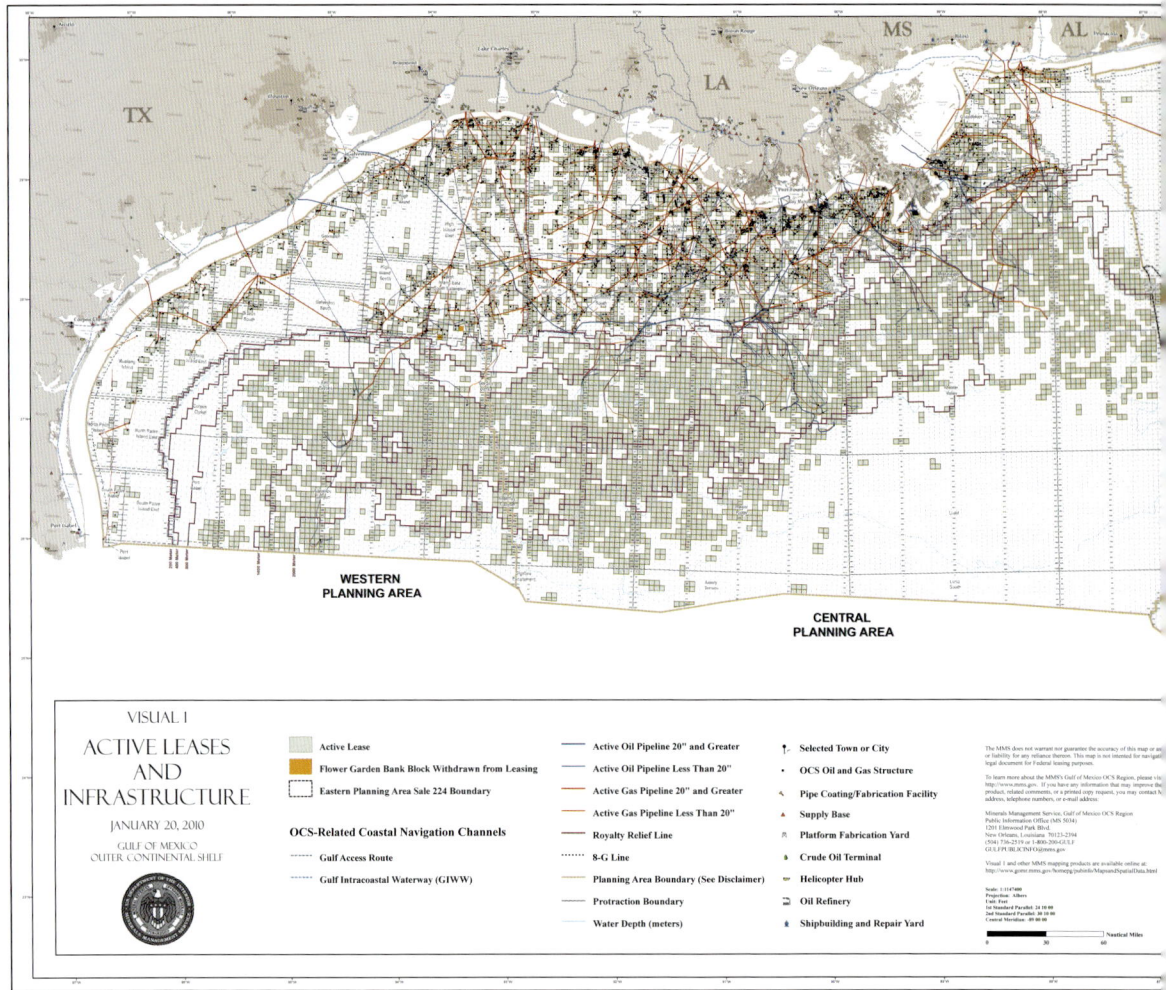

Map of oil leases in the northern Gulf of Mexico. Note that oil and gas production for the most part is prohibited off the west coast of Florida. Map courtesy of US Bureau of Ocean Energy Management.

environmental impact statement follows. BOEM informs the public that an area is under consideration for oil and natural gas leasing in the *Federal Register*, and invites public comments and input from interested parties. An initial examination of the area allows the government to consider the value of energy exploration versus potential environment harm or conflict with other uses of the seabed (for example, fisheries).

The next step, often the longest, is the production of an environmental impact statement (EIS).

The EIS, commissioned by a bidding company or by BOEM, can be conducted by an environmental consulting company, by a team from one or more universities, or by a combination of contractors. The EIS must consider what resources may be present and how they may be developed, and what alternatives to exploration exist. The EIS requires a detailed description of the existing environment, an analysis of possible impacts on the environment, an accounting of any unavoidable adverse environmental effects of exploration

EASTERN
PLANNING AREA

MMS U.S. Department of the Interior
Minerals Management Service
Gulf of Mexico OCS Region

information from the final review committee are incorporated into the final EIS.

Unless new information reveals other potential detrimental environmental effects, the bidding process proceeds to a "Proposed Notice of Sale." The offices of the governors of the affected states now have a chance to comment on the proposed sale. If there is no particular opposition to the sale, it goes to the "Final Notice of the Sale." Companies submit their bids for the blocks they want to explore. The bid package includes descriptions of planned exploratory work, as well as the amount of money the company is "exposing" to obtain access to that particular block. The highest successful bid in a 2009 round in the Gulf of Mexico was $28,133,843. The block goes to the apparent high bidder, after which leases are issued. The BOEM may decline to sell a lease if no bidder offers what is considered fair-market value for the area. No guarantee is given that a lease will be profitable, and the successful bidder does not have to drill in the lease immediately. Lease areas consist of nine square miles (three miles on a side) and are good for from five to ten years.

Following the lease sale, it may take several years for the winning company to actually drill an exploratory well on a leased block. Sometimes a company will fail to carry out a minimum level of exploratory work on a block and may relinquish the block to the government for a later leasing round. When the company does carry out exploratory work, it typically takes the form of seismic and drilling activity. If the seismic data is promising, an oil company will invest many millions of dollars to drill an exploratory well. If the initial well is promising, appraisal drilling may follow while the company works to determine whether a full development project is warranted to produce the oil, gas, or hydrocarbons discovered at the site.

and extraction and the potential mitigating measures that could be taken, and a full supplement of supporting documentation. Special attention is paid to rare, endangered, or commercial species, such as deepwater corals, sea turtles, whales, and reef fishes. Thirty days after publication of the draft EIS in the *Federal Register*, but within a 69-day comment period, one or more public hearings in the vicinity of the proposed lease sale area are held (in Texas, "in the vicinity" often means in New Orleans). The comments received and any

2

RIGS AND
PLATFORMS

MOST PEOPLE speak of all oil and gas production facilities as "oil rigs." But to be technically correct, a petroleum production facility in the process of being installed is called a "rig," distinguished from an established facility, called a platform, by a drilling tower. In short, a rig drills, a platform produces. At any particular time, one is more likely to see platforms, not rigs. Fishing or diving is done at a platform, not a rig.

The characteristic tower of a drilling rig can signal an early stage in the history of a future platform. When drilling has located oil or gas, the extracted oil or gas is tested to determine its consistency and volume and to predict whether or not a particular rock deposit will yield enough product (oil, gas, or hydrocarbon) to warrant continuing work at the site. Not every drill site will be worth further effort. The probability of success depends on the quality of the seismic data and its interpretation, and on how much oil or gas actually exists in any given area, how much exploration already has occurred in the area, and how much risk a

company is willing to assume, among many other factors.

Some drilling is carried out by rigs secured to the seafloor by legs, as is the case in jack-up rigs. Other drilling operations, such as those in deeper water, occur from the floating platforms of drillships, semi-submersible platforms, and other specially designed units. These may be moored or held in place by dynamic positioning technology, which uses thrusters and sophisticated sensors to keep the vessels on location. The drillship contains a well bay through which the drilling pipe goes down.

Jack-up rigs, often seen in the Gulf of Mexico, can be towed to the place of operation. En route, the legs and main structure are jacked up so that the rig floats; on site, the legs are jacked back down. The legs have large pads that do not sink far into the sea floor. The legs can be lengthened or shortened by taking off or putting on extensions on the upper end to allow it to operate in different water depths.

Rigs also can be semi-submersible: the rig is

A jack-up rig is held in place by legs that are jacked down to push the rig out of the water. The high tower is the "Christmas tree," used during drilling and characteristic of a rig. Photo by Dick Zingula.

on a huge float with no legs and is held in place by anchors when on location. It is fairly common to see these rigs waiting to be towed or being repaired in Texas harbors such as Galveston, Freeport, Port Aransas, and Ingleside.

If the drilling turns up a deposit that is worthwhile, a company can then extract the oil or gas through that well. The system can be as simple as placing a pump on the wellhead, or as complex as installing a multi-million dollar development consisting of a floating platform and multiple wells. Either way, pressure pushes the oil up through the well. A single platform can be used as a base of operation for up to thirty wellheads located at different water depths and at remote locations up to many miles away from the platform. Modern drilling technology enables a company to drill not only straight down, but also diagonally or at different angles. Remote wellheads may be attached to a platform by means of a distant umbilical connection, which operates like a power cable to operate the equipment, while a flow line carries the production from the wellhead to the platform.

The type of platform depends on the depth and local conditions of the water, including currents, waves and swell, and an assessment of how severe storms can affect those local conditions; on the characteristics of the reservoir; and on operating costs and operator preference. Nearshore platforms in the Gulf of Mexico are often fixed platforms anchored directly into the seabed on a framework of steel legs and crossbars commonly called the "jacket." The well shaft proper extends vertically down from the deck to the sea

A semi-submersible rig has no legs. It is held in place by anchors below the surface. Photo by Dick Zingula.

floor and the submerged deposits. Diffuser pipes for gasses and saltwater intake pipes also extend under water. Deepwater platforms (at more than 1000 feet or 303 meters deep) may be attached directly to the seafloor or can be floating types constructed of steel, concrete, or steel/concrete composites. The floating decking and other structures are moored to the seafloor and connected via pipelines or risers to the wells far below the surface of the water. Pipes can connect a single platform to multiple wells, and platforms can occur in linked groups or singly. Platforms can be named according to an abbreviation of the leas-

ing tract (for example, "HI," meaning High Island, and then a number or letter combination, 389A, to indicate a particular structure), in alphabetical order (Charlie, Lima, etc.), or according to a nearby geographical feature (Little Adam is near Big Adam Bank, for example).

Platforms can be always manned, occasionally manned, or never manned. Unmanned platforms consist of a well bay and a small lighted surface pad for maintenance. Some unmanned platforms may also have an emergency helipad and shelter. All platforms have a means by which maintenance ships can be secured to them, and they are all equipped with lights that serve as navigation aids. Some are equipped with loud horns or whistles that also warn ships of their location.

A manned platform is much more complex. At the surface, there are emergency ladders and usually large floating fenders that protect the jacket from the supply boat. A series of steps connects the landing dock to the main deck proper, but usually people and supplies can be loaded in a lift basket from a loading boom. As a general rule, supply ships deliver heavy supplies such as pipes, whereas personnel and light cargo come by boat or by helicopter.

Much of the structure of a platform that is above the surface of the water is dedicated to oil and/or gas production, including monitoring of the oil extraction process and movement of the extracted oil into delivery systems. Platforms are amply supplied with lights to aid 24-hour operations aboard as well as to provide a navigational warning.

Flare booms, which safely vent ignited natural gas, are visible on some platforms, especially at night. If there is an easy, inexpensive way to transport the natural gas to market, the company will do so. If not, the company will often flare it off because natural gas is considered less valuable than oil. A more cost-effective way to dispose of

Platforms can occur singly (A) or in pairs (B). Notice all the fishing boats around the large platform. Photos by Dick Zingula.

natural gas produced during pumping is to inject it back into the reservoir to boost or maintain the reservoir pressure, which could increase production levels.

Crew quarters on oil platforms tend to be relatively spartan. Often there are two shifts per day for a crew of at least a dozen people, each with specific skills. Many of the personnel are transferred on and off the platform by helicopter for two-week shifts. Most oil companies realize the value of good food for their crews and employ cooks. Newer offshore facilities try to offer a homier feel for their residents by having gyms, entertainment areas, and internet access.

A platform and its parts. Living quarters and the main operational facilities are topside. The large supporting beams and their connectors are called the jacket. The main conductors, which collect oil and gas, lie in the center of the jacket. Photo by Jesse Cancelmo.

NATURAL REEFS AND BANKS COMPARED TO THE PLATFORMS

PLATFORMS HAVE been described as artificial reefs. A single platform can introduce up to three acres (8,000–12,000 square meters) of surface area above the seafloor. Like a reef, it introduces vertical relief above a more-or-less flat, monotonous seafloor. The platform provides an attachment site for seaweeds and marine invertebrates, as well as a hiding place, feeding site, and nursery area for fishes. But a platform differs from a natural reef in many ways. The framework of metal alone discourages the colonization of some marine algae and animal species that prefer to settle on natural substrate such as rocks or shells. Much of any platform consists of vertical elements, often with additional crossbars and angles that provide limited horizontal surfaces. None of the natural reefs of the western Gulf break the surface: the shallowest depth at the East Flower Garden Bank is 65 feet (20 meters), while platforms extend well above the waves.

Platforms located within a few miles of natural coral reefs differ considerably in their in-habitants and topography. A platform does not have large coral mounds and pillars. There are no sand patches on the platform, and thus no sand-dwellers such as southern rays, jawfishes, conchs, or swimming crabs (family Portunidae). Various clams can bore into soft rock or dead coral of natural reefs and make holes that can be inhabited by a variety of small worms, crabs, and shrimps. These burrowers are absent on the metal platforms. Certain shade-loving sponges, nocturnal species such as cardinalfishes and hinge-beak shrimp, and morays are largely absent, probably because there are no holes on the platform deep enough to satisfy their need for shade and shelter.

The "fishing banks" located offshore of Mustang Island and Padre Island on the central and southern Texas coast consist of soft, easily broken rocks. Boring clams and species that seek shelter in cracks live on the banks and not on nearby platforms. These banks, like nearby platforms, often are coated with sponges, tube-building worms, and sea whips. Biodiversity on these banks as well

A complex framework of legs and beams supports a platform and creates a habitat for fishes and other marine life. Photo by Dick Zingula.

as platforms in this area is severely impacted by the nepheloid layer: an area of silty water that pours out of the Laguna Madre and rivers from the Brazos to the Rio Grande. Clay minerals in the river water may have a slight electrical charge that prevents them from settling, while fine particles of silt, decaying plant material, and debris can be easily resuspended by surf. Especially in summer, the nepheloid layer tends to sink to 60 feet (20 meters) and below, coating attached animals and clogging their feeding mechanisms and gills.

It is in this area that platforms have an advantage over natural banks as a habitat because they extend above the nepheloid layer. The upper areas of production platforms remain in relatively clear water. Large sponges and dense aggregations of pink barnacles, which otherwise would be smothered or be unable to settle in silt, thrive on the platforms.

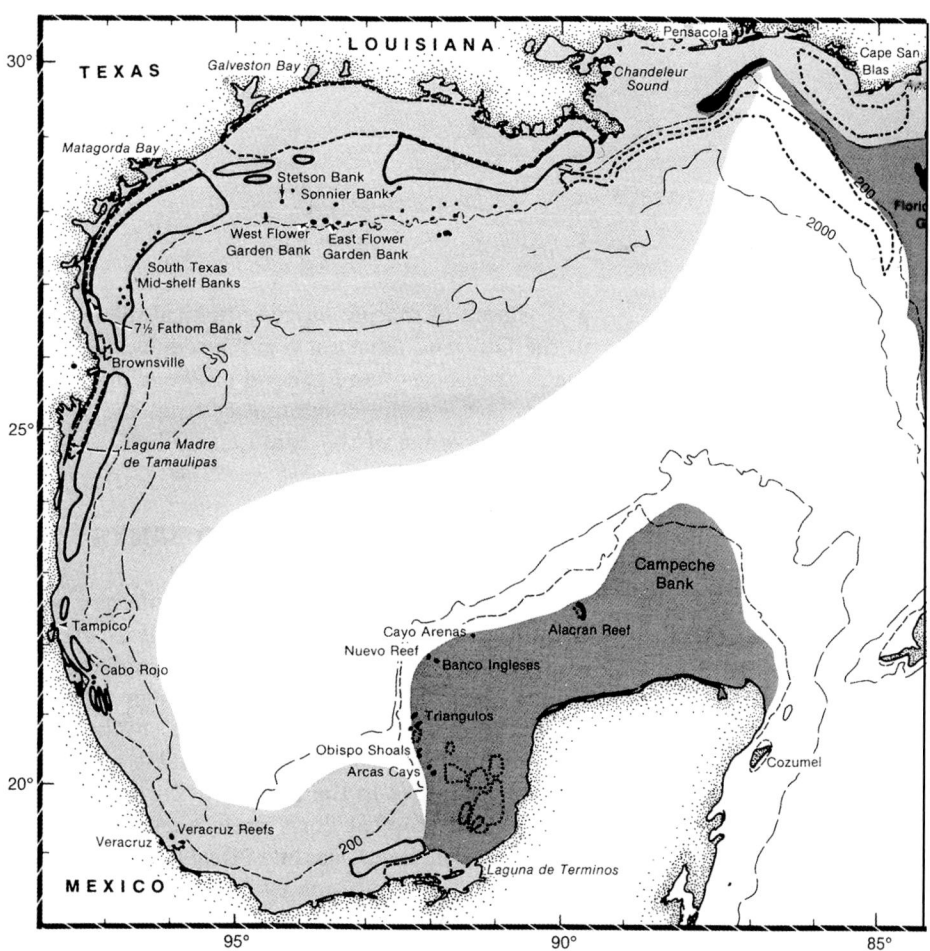

Much of the coast of the northern Gulf of Mexico consists of mud, but there are isolated hard banks offshore. The coasts of Florida and the Yucatan Peninsula consist of hard limestone and support more extensive shallow-water reefs. Map from Rezak et al. (1985).

4

HOW LIFE
ARRIVES AT THE
PLATFORMS

THE FINISHING touches are done; the warning lights function and the waves lap at the base of the jacket. A crew is working on the deck far above the water. Organisms of land and sea now can either colonize (establish permanent settlement) or visit (take up short-term residence).

Visitors above the Waves

Among the visitors will be many of the common coastal birds, such as laughing gulls, which follow fishing boats offshore and may stop to rest on platforms. Caspian terns can fly out to 70 miles from shore to forage, and often pass by the platforms. On some platforms, models of owls have been installed in hopes of discouraging birds and their inevitable droppings. Peregrine falcons have been known to stay on platforms, where they hunt seabirds and the migrants that land there. During spring, hundreds of black terns and brown boobies fly by the platforms, while magnificent frigate birds show up in summer. Near platforms out at the edge of the continental shelf, pelagic birds such as shearwaters may soar over the waves.

Every year, hundreds of thousands to millions of birds migrate across the Gulf of Mexico between the southern United States and Mexico or

Black terns migrate in great numbers across the Gulf of Mexico, and pass by the platforms. Photo by Terry Sohl.

14

Laughing gulls are among the most common birds year-round in the northwestern Gulf of Mexico. These birds often follow boats along shore all the way out to mid-shelf platforms.

beyond as they fly south during the fall and return in the spring to breed on the mainland of the United States and Canada. Each flight can take up to 18 hours of steady movement. Flying across the Gulf of Mexico, birds have no access to food or fresh water. A headwind can cause birds to quickly burn through their energy reserves, a storm can slam birds into structures above the water, or capture them in a swirl of turbulence around a platform. A short stop on a platform, especially one that has some puddles of rainwater, can be a welcome and even lifesaving break. Much like on the shore after a storm, when a large number of birds arrive at once, the phenomenon is called a fallout. Many of these small flyers will not survive the migration. It is a sad sight to see a waterlogged small songbird floating just below a platform, or watch it disappear into the rising swirl made by a large fish.

Fifty species of birds have been reported at HI A389-A platform, which is located within the boundaries of the Flower Gardens Banks National Marine Sanctuary. Many of these birds are regular migrants across the Gulf of Mexico, but surprising visitors can reach platforms after storms.

Insects, too, may visit the platforms. Butterflies may migrate over the Gulf. They flutter close to the waves, getting some lift from the updraft between crests. Dragonflies, moths, flies, ladybird beetles, and other insects may "hitchhike" on boats or be carried offshore by wind and eventually move to a platform. With no food available, these insects do not survive for long, but may provide a welcome snack for a migrating bird or a hungry fish.

Whether migrating across the Gulf of Mexico or blown offshore, land-dwelling animals can come to rest on the platforms. (A) Goatweed butterfly, *Anaea andria*. (B) The pigeon stayed aboard as a pet for weeks until a boat finally took it back to land.

Below the Water
Recruitment and Succession

The process by which living things move into and inhabit an area is called recruitment. Some species never recruit to a platform, or if they do, they do not survive. The amount of light can be a major problem for many types of algae, whether one-celled or multicellular. At any given time, only one part of the platform will be exposed to sunlight, with much of the jacket shaded by the deck above. Seaweeds require adequate light to survive, so many species cannot live in the shade of a platform. Near shore, where the water often is full of silt, only a few species of algae can inhabit the platform in areas near the surface of the water.

Many reef corals cannot survive on platforms because they interact with microscopic algae, called zooxanthellae, in their tissues. The zooxanthellae require adequate light and rarely withstand water temperatures of less than 65°F

(18°C). If these algae cannot survive, neither can the corals. Even platforms within one mile of a natural coral reef have very limited coral growth. For a massive coral such as those found on the Flower Gardens Banks, assuming a growth rate of at most 0.5 inches per year, it would take at least four years before the coral colony would be readily visible to a diver.

Upper areas of any platform can be swept and scoured by waves and currents, preventing the settlement of animals such as periwinkle snails or small algae that usually live in areas exposed by low tide. If there are strong bottom currents below a platform, turbulence can scour away any sand. This scouring leaves a seafloor of shells, rock, or coarse sand that scrapes the legs of the platform clean. Outfall pipes from the platform contain produced water (wastewater from oil production). By law, contaminants and other substances must be removed from the water before it is discharged back into into the Gulf, but the flow associated with the discharge may scour the nearby seafloor.

If there are no strong bottom currents, the seafloor below any platform usually consists of soft sand or mud. This seafloor is populated with many small organisms, especially segmented worms and small crustaceans. Although these widespread animals are likely to be found below most platforms, they do not colonize the platform itself. In some areas, biotic communities of the seafloor below the platform and the biota of the platform itself may have no species in common.

The majority of the settlers on the platform itself arrive as spores or larval stages carried to the platform in the water. Some fishes, shrimp, and large crabs can travel across the bottom between platforms, but they are in the minority. Most of the animals living on platforms would either sink into the sediment or be quickly picked off by predators if they left the hard surface of the platform. Even if they could cross the seafloor, animals living on the platforms can rarely climb from the bottom to shallow depths, a vertical climb of more than 100 feet. Many animals, especially corals and sponges, have almost no mobility at all. On deeper platforms, many animals are confined to certain ranges of depth by changes in pressure and temperature, and cannot survive above or below certain limits.

After a platform is installed, the seafloor immediately below it must recover from disturbance: the drilling itself, disposal of drilling mud, installation of the legs and other equipment, and the creation of small eddies or scoured areas as currents pass around the bases of the legs and other hard objects. Recruitment of new animals here may be fairly rapid because bottom-dwellers are able to swim or crawl across the sediments. Even so, it often takes years for local communi-

ties of small clams, worms, crustaceans, and burrowing sea anemones to reestablish on the sea floor. Small-scale patterns of shelter or exposure to near-bottom currents often result in patches of different species across scales as small as a few feet or less.

Recruitment of larvae often proceeds from the south, then west to the east, following the common pattern of currents in the northern Gulf of Mexico. Currents enter the Gulf between Cuba and the Yucatan Peninsula of Mexico. Flowing northward, the most prominent current is the often strong Loop Current, which moves northward toward Alabama, loops eastward, and then heads back to the south, exiting through the Straits of Florida. Maps of the currents in the western Gulf of Mexico generally show two patterns. In the first, one stream passes northwestward from the main flow of the Loop Current and then flows east to west from the area of the Mississippi Delta. In the second, another stream flows from south to north along the shoreline of the western Gulf of Mexico or in the waters well offshore of the area. Both of these streams tend to converge in the vicinity of Big Shell on Padre Island National Seashore. From there, they may join in a subsurface current that eventually exits through the Straits of Florida. But this pattern is a generalization. Drift bottles released into the Loop Current or off the coast of Texas can show a more or less westerly flow or a northerly one depending on the wind direction. The ocean current website of the National Oceanographic and Atmospheric Administration (NOAA) gives further information from buoys and satellite data on ocean temperatures, current flows, and general information on the Loop Current and other ocean currents.

Circular streams called gyres break off from the main currents or form in local areas. These gyres play an important role in the transport of

Map of currents in the Gulf of Mexico. From Britton and Morton (1989).

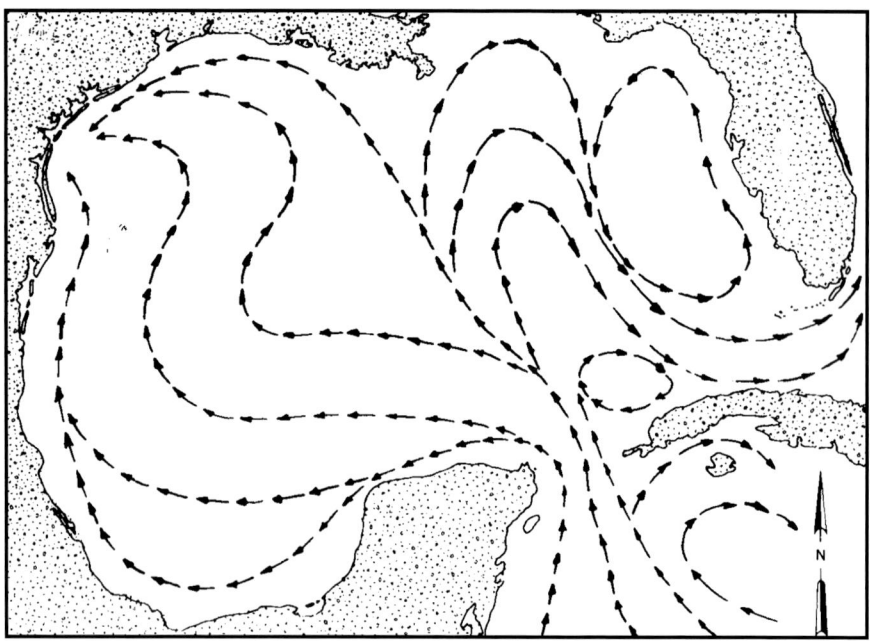

heat in the Gulf as well as salinity regimes and movement of nutrients. Larvae that become caught in a gyre may end up well away from the main flow of the Loop Current. Year-to-year variability in the number and species composition of larvae must be expected. Genetic comparisons between different populations can give clues to their relationships, but so far, few studies have compared organisms on the platforms with their relatives on natural hard surfaces. Such studies may detect the ancestral stock of marine life from which the isolated animals and algae of the platforms have been derived.

The common algae and animals that settle on platforms generally inhabit hard surfaces: rocks, coral skeletons, or shells. Before human activity, such surfaces were in short supply along the coast of the northern Gulf of Mexico. Natural hard surfaces included oyster reefs in very shallow water, some submerged fossil oyster banks, the northern Texas-Louisiana reefs, and the southern Texas fishing banks. Unlike in southern Florida or farther

south along the Mexican coast, none of the reefs reach the surface. The nearest rocky shores are in southern Florida and in the southern Gulf of Mexico.

An element of chance is always present in colonization of an area distant from shore or a natural reef. Not everything that ends up at a platform may be able to settle there. Studies of fish larvae collected near platforms show that an enormous variety of species arrive there, including larvae of many species that live on the muddy seafloor. Even if a species can settle there, platforms very close to one another can differ markedly in their species composition. The length of time that larvae can survive in the water column before they must settle to the bottom varies by species. Some shrimp larvae can last up to six months in the water, but many larval tunicates must settle within forty-eight hours. Pushed along by hurricane-force winds, the most unlikely larvae can end up in the northern Gulf of Mexico. A single flaming reef lobster

(*Enoplometopus antillensis*), for example, turned up at Stetson Bank, although this species rarely occurs north of the Caribbean. Many species reproduce seasonally and their larvae may not be present in the water at the time a platform is ready for settlement. The first lucky settlers may be able to crowd out or overgrow latecomers, or move into an empty space caused by damage from a storm. The arrival of new settlers dependent on chance is termed "sweepstakes dispersal."

Soon after any large object is placed in the water away from shore, it may become a haven for small fishes and juvenile crabs. Bermuda chubs and "baitfish" such as sardines, juvenile jacks, and sergeant majors often arrive quickly and use the platform as shelter. More commonly, the first set-tlers on a platform are slimy microscopic films of bacteria and other one-celled organisms. Within twenty-four hours, populations can reach several thousand individuals in less than a single square inch. In time, these organisms give way to a cov-ering of algae or hydroids, tiny relatives of sea anemones. The stalk-like projections of the hy-droids provide shelter for amphipods, which are tiny shrimp-like crustaceans, and various marine worms.

Jack-up rigs tend to be colonized first by bac-teria and small fishes. The areas near the sur-face on these rigs, as well as the floats of semi-submersible rigs, often are settled by gooseneck barnacles (*Lepas anatifera*). A leathery, movable stalk to the nearby metal attaches these odd creatures. They use feathery legs to capture small floating food particles from the water. They com-monly are found on floating objects, such as drift-wood, fishing floats, discarded styrofoam, and other "junk." Within six months, they can form a dense covering. Life for these animals is limited to areas near the surface, for at greater depths they quickly are crunched and eaten by fishes, crabs, or sea turtles. Unable to withstand drying, when the

Hydroids are early colonists of near-shore platforms. At a distance, hydroids look like algae (A), but look more closely (B). A set of tentacles similar to those of a sea anemone surrounds the mouth of each animal. Each animal is part of a colony set in a flexible bushy covering. Photo "B" by Dick Zingula.

legs of a jack-up rig are raised, gooseneck barnacles dry out and die.

The earliest settlers on a fixed platform often are microscopic films of algae, hydroids, and sponges. These species spread from one firm surface to another as spores or larvae, carried by the ocean currents. Upon settling, they undergo metamorphosis—a change from larval or juvenile forms into the adult stage.

Given a few months, the early settlers are joined by the larvae of sessile barnacles (without stalks but cemented directly to the metal). Barnacles soon will become the dominant attached animals on the platform, in both areal coverage and wet weight. The well-studied life cycle of a barnacle provides a good example of metamorphosis.

Once hatched from the egg, the larval stage of a barnacle is called a nauplius. It looks like a stubby, one-eyed shrimp. After growing larger and adding new legs, it changes into a cypris, a stage in which the larvae becomes enclosed in a two-part covering and is ready to settle onto the platform. Using its antennae, the cypris tests a surface for roughness, silt, presence of adults of its species, and other environmental features. For many barnacles, presence of adults is the most important criterion that institutes settlement. On a bare surface, the sensory structures and the simple central nervous system of the cypris seem to combine to evaluate the surface. A cypris can lift off and resettle several times, but most barnacle larvae must settle within a few days after becoming a cypris. The animal secretes biological glue and cements itself to the surface. The covering is shed and the animal grows a series of plates. It now is an adult barnacle.

One barnacle attracts another. Within a year, barnacles are all over the upper parts of the platform. Their closely packed plates create a three-dimensional surface that becomes a site for hiding or attachment of additional species, such as tiny crustaceans and worms. These small animals attract small predatory crabs, fishes, and worms, which in turn provide food for bigger fishes such as snappers. Sponges may arrive as larvae and then overgrow the barnacles or colonize bare spaces. Eventually, the big predators—the groupers, amberjacks, and king mackerel—find the platform and prey on the smaller fishes.

The gradual change in faunal composition at a given location over time is called succession. The exact pattern varies with how far offshore the platform is located, when it is erected, and any measures taken by the platform's operators to inhibit settlement and growth of marine life. Near shore, where seasonal pulses of low-salinity water and heavy silt smother platforms in winter, there may be die-offs of species every year and replacement of these animals during the next spring.

The arrival and eventual settlement of a platform by seaweed and animals is a mixed blessing. On one hand, the enhancement of local fishing by providing food and hiding places for fishes can benefit the local environment. On the other hand, massive growths of marine organisms can corrode the underlying steel. Barnacles can cause severe local pitting of steel and may even gouge their way through coal tar coatings. The water that becomes trapped under mats of micro-algae may become acidic. Some operators apply anti-fouling paints (paints that inhibit settlement of marine life on a surface) or copper/nickel cladding and use corrosion-resistant materials, combined with periodic inspection and scraping, to prevent excessive buildup of attached organisms.

Numbers of Species

Oil platforms offer an ideal place to test one of the most interesting questions in ecology, that

Gooseneck barnacles are early colonists of some platforms but also commonly settle on driftwood, styrofoam, and other floating objects.

Barnacles such as *Balanus reticulatus* (pink, striped), *Balanus eburneus* (white), and other species are common early settlers on nearshore platforms. Each barnacle in the photo is about the size of a large pea.

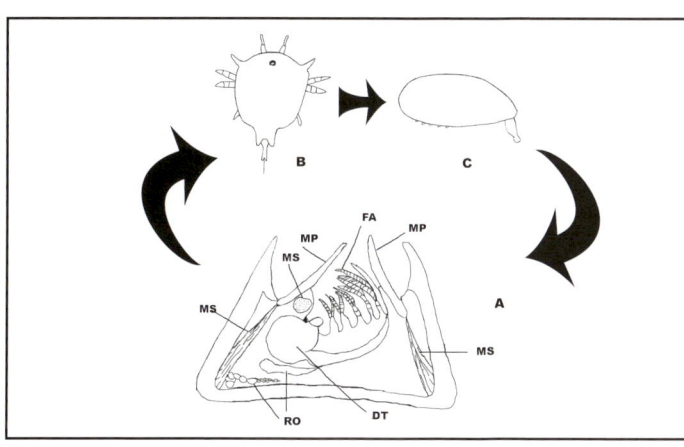

Life cycle of a barnacle. An adult barnacle (A) releases the first larval stage, the nauplius (B). The nauplius goes through several molts before becoming the settling phase, the cypris (C). Using glands in the antennae, the cypris cements itself to a surface and then changes into the adult. Parts of the adult: DT= digestive tract; FA= feeding appendages; MP= movable plate; MS = muscle; RO = reproductive organ.

of species packing: the number of species that can occur in an area. An area where environment conditions are relatively stable year-round is more likely to have the most species. In a stable environment, specialists are found: animals that have a particular relationship with other species, such as small snapping shrimps that only live among the chambers of sponges, or organisms that prey only on selected food items, such as the snails that feed only on the tissues of sea whips. A disturbed area such as a silty, nearshore environment will have fewer species.

It can be very difficult to estimate the number of species that will settle on or visit a given platform. A platform constitutes a habitat different from that of the surrounding seafloor and isolated from other habitats; therefore, it can be considered to be a sort of island. In their book *The Theory of Island Biogeography* (1967), ecologists Robert MacArthur and E. O. Wilson proposed a formula to estimate the number of species that would be present at equilibrium. This point would occur when the number of new species arriving at the platform per unit of time (the immigration rate) was balanced by the number of species already there that were leaving, either by emigration or death (the local extinction rate). At first, many new species can arrive on the unoccupied platform. As time passes, the local extinction rate increases because settling species cannot find adequate food, are overgrown by barnacles, are eaten by predators, are unable to reproduce, or are otherwise adversely affected. Note that "extinction" here refers to a given platform, not the entire world.

The starting point for immigration is the number of potential recruit species from the source of recruits—a natural reef, another oil platform, etc. These species move on their own or are carried as larvae, spores, or fragments away from the parent population. Not everything in a source area can be an immigrant. Seagrasses, for example, are common along the coast of Texas, but cannot settle on the vertical surface area of a platform. The end point for extinction at a platform is theoretically equal to the starting number of potential immigrants.

Matters are not so simple in the real world. The size of a platform and its distance from a source of recruits are important. The bigger a platform, the more area it has for species to settle. The further it is from a source of recruits, the fewer species exist that can cross the distance needed to settle there. Thus, plots of the intersection points of immigration and extinction will lie at different numbers of species for different platforms.

To add other factors that affect the number of species, more than three dimensions would be needed. An older platform is likely to have more species than a newer one because it takes time for larvae to settle and fish to find the place. A structurally complex platform with three or four legs has more surface area than a platform consisting of a single large spar, and thus more room for species. Then there is the Archipelago Effect—it is easier to recruit from one island in a chain to the next island than to cross a long distance between a source of recruits and an isolated platform. If other platforms are located up current from the platform in question, recruitment may be much faster than it would be from a natural reef far away.

An equilibrium number is an ideal situation. A platform located over 50 miles from shore may reach equilibrium in a decade. A platform near shore, where storms and silt frequently move through, probably never reaches equilibrium so species numbers continually fluctuate, as do the numbers of species on a platform that is regularly scraped clean by divers.

A species saturation curve number can be

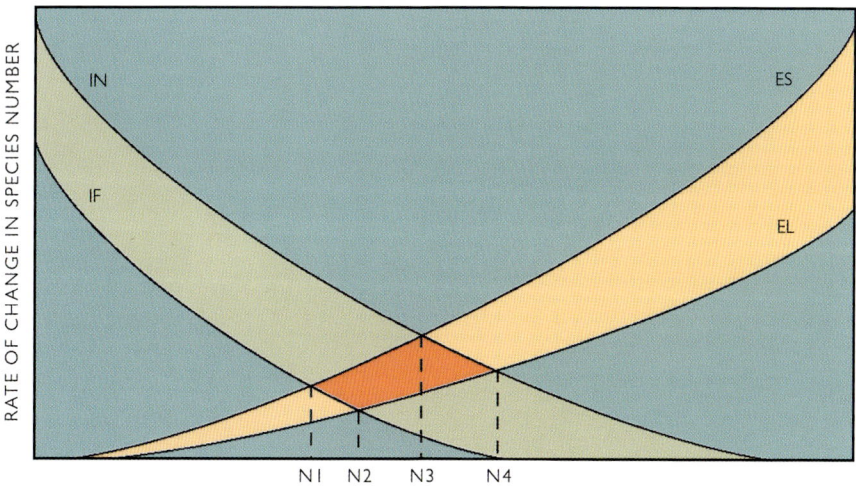

Hypothetical MacArthur-Wilson curves for rates of change in species numbers and numbers of species at equilibrium for different platforms. IN= immigration rate for a platform near to a source of recruits; IF= immigration rate for a distant platform; ES= extinction rate for a small platform; EL = extinction rate for a large platform. NI–N4 are equilibrium species numbers for different platforms.

used to determine if the number of species is stabilizing, in which the number of new species encountered is plotted against the passage of time. Usually, the number of species recorded at a platform increases sharply at first: at the end of the first year, a diver may see only four or five species; by the end of the second year, another five species, in additional to the original species will be seen, and so on. Eventually, few species new to the platform will be seen. The curve flattens and equilibrium has been reached. In extremely diverse areas such as the coral reefs of Indonesia, every time a biologist goes to the study area, he is likely to find something new, so the curve keeps on rising for years or even decades on end. In a badly polluted environment, the curve flattens.

To determine the settling rate of new species on platforms and other man-made structures settling plates are used. These ceramic plates are left to "soak" on the site for various lengths of time, and then the contents are compared. These plates

help to determine the sequence of settlement up to barnacles. The small size of these plates (generally no more than one square foot) prevents the biologist from following the pattern of change on blue-water platforms where huge sponges or coral colonies may settle. The resident sponges, barnacles, and other larger invertebrate animals settle on the platform within five years of placement, but they may grow very large over time.

Estimating the growth rates of animals on a platform can be very difficult, because most species exhibit no signs of regular growth patterns. How long a particular sponge, colony of barnacles, or coral will live also is unknown. Platform HI A389-A, within the Flower Gardens Banks Natural Marine Sanctuary, contains some bucket-sized sponges and clusters of orange cup coral, some of which could be over 25 years old. Evidenced by the piles of barnacle shells and bits of dead coral beneath the platforms, as well as by drill holes and crunch marks made by predators,

Species saturation curve. As time passes, the number of species discovered in an area eventually levels off. For this hypothetical area, the equilibrium number of species would be about 20.

some natural mortality occurs. Theoretically, the biota of the platforms has a starting point of 1938, the year in which the first platform was erected off Louisiana. But no individual platform has gone undisturbed since that date. Since then, many platforms have been moved, slammed by severe storms, continually scraped, or taken out of operation.

The exact number of species for any single platform is unknown. The estimated number of encrusting species and fishes on a platform just off the coast of Louisiana is fifty species or less. About 27 miles offshore of Galveston or Freeport, Texas, 101 species of invertebrates have been reported from a platform in the Buccaneer Field. Platforms located in clear water over 50 miles offshore, where there is less marked seasonal change and greater vertical depth, may be inhabited by hundreds of species. If a fast-growing tunicate or soft coral colonizes the platform, the invasive species may crowd out or overgrow other species and "take over," thus reducing the number of species in the area.

Once a platform is in place, the species population will shift from that of the natural muddy seafloor to that of a hard surface. Species such as commercial shrimp, croakers, and crabs will lose habitat. Sponges, snappers, and barnacles will be the winners. Whether this shift is viewed as "good" or "bad" depends on economic and aesthetic considerations. Endangered or threatened species such as sea turtles and peregrine falcons may visit platforms, but they do not nest there, nor do corals establish reef structure on platforms.

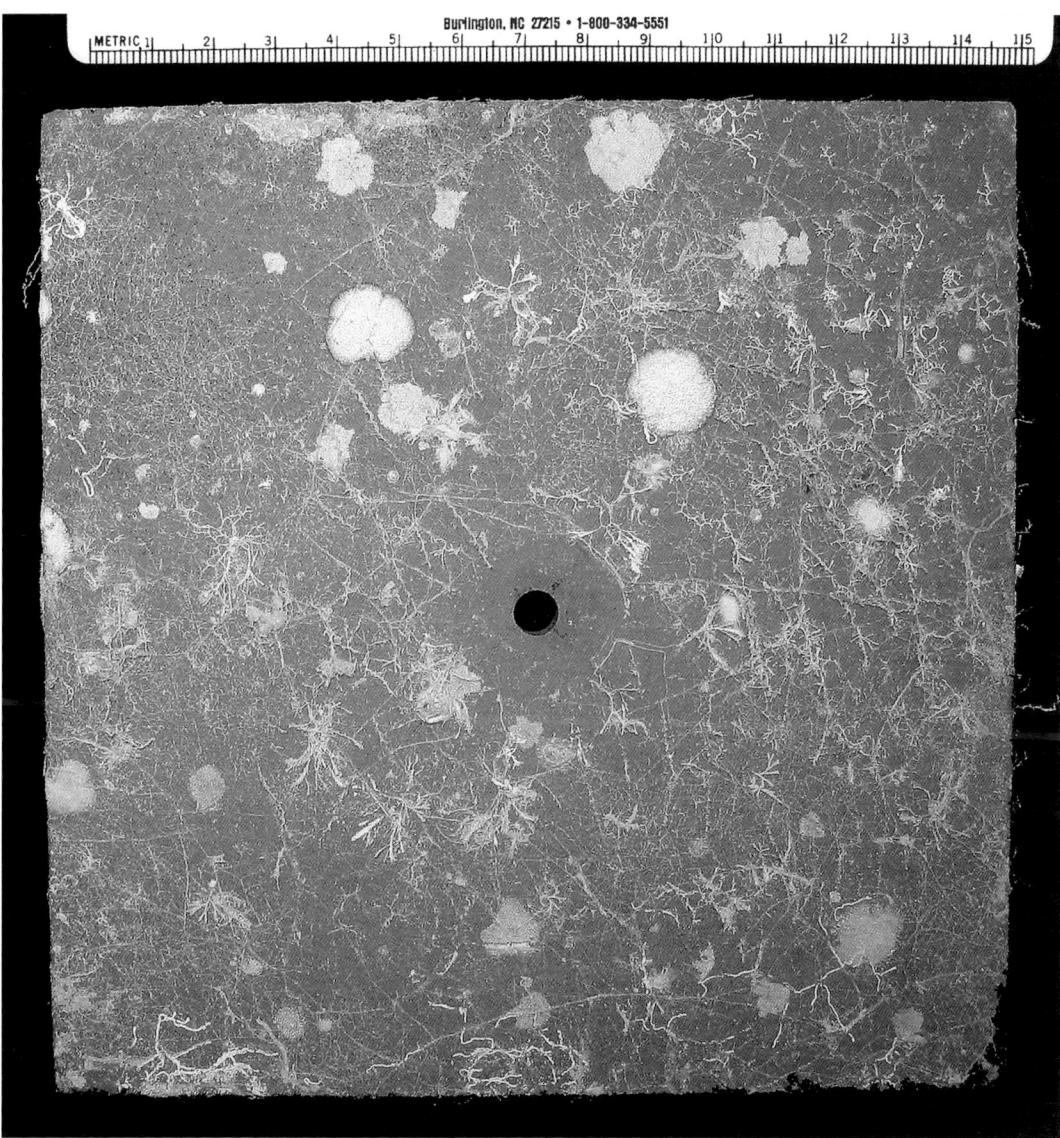

A settling plate is bolted to a surface and left for a period of time before it is removed for examination. This plate, "soaked" for three months, had been colonized by bryozoans (flat patches) and hydroids (thin bushy tubes). Photo by Archie Ammons.

5

▼

BIOLOGICAL
INVASIONS AND
PLATFORMS

A BIOLOGICAL invasion occurs when non-native species move into new habitats where there are few or no checks (e.g., predators) on its population growth so it spreads and reproduces rapidly. The effects of such an invasion can be difficult to determine. Often, the introduction of a rapidly reproducing "weed" species can be economically or ecologically damaging. Introduced species can outcompete or overgrow native species directly, or may feed on larvae, serve as a host for parasites, or produce noxious chemicals. A common indirect effect when an invasion occurs in the ocean is that the food supply for bigger fishes is reduced because the smaller animals on which they feed are crowded out, poisoned, or eaten. One of the worst marine invasions is currently taking place in the Mediterranean Sea along the coast of France. Acres of seafloor have been overgrown by the green alga *Caulerpa*, which is inedible to most marine animals and covers over or shades out local native algae and corals, sponges, and other attached invertebrates. This seaweed seems to

have accidentally been released from aquaria and is spreading rapidly via fragments that break off and attach to the seafloor.

With modern ship traffic and trade in live seafood and bait, marine animals now are introduced into new areas, intentionally or not. Some of these introductions may be beneficial: for example, the deliberate introduction of sea turtles from Mexico into Texas to enhance a breeding population. Others are intentional but ill-considered, such as releasing overgrown exotic aquarium fishes on reefs off the Florida Keys. Many are unintentional. To maintain proper flotation, ships partially flood their holds with seawater. This ballast water is usually neither filtered nor treated to remove any larvae or other small organisms. Enclosed in a vast "aquarium," many of these small animals can cross the entire Pacific Ocean or even go through the Panama Canal. At the next port, the ballast water may be flushed and the enclosed organisms liberated. As a result, many bays, rivers and coastal river mouths have been infested with exotic animals, including parasite-laden crabs, overgrowing tunicates, and larvae-eating jellyfishes. Some states now have passed laws requiring oceangoing ships to either treat their ballast water or discharge it out at sea, where the trapped organisms are less likely to survive. But not all states or nations have such laws or the means to enforce them, allowing currents to spread larval stages of transoceanic species to still other areas.

Exotic species also may be attached to a hull or the jacket of a platform if it has rested in a bay for more than a month. Such attached animals and algae are called fouling organisms. Today, ships entering the water of Texas must be "scamped"— their ballast water must be treated with compounds that kill larval stages. But such regulations are new and may apply only to vessels in the Gulf of Mexico. Organisms may have been introduced into the Gulf of Mexico decades ago, but with no baseline studies, it can be impossible to determine when a species first arrived. It can be very difficult to determine which fouling organisms are truly native to the Gulf of Mexico. Many of these species now occur worldwide near harbors, the result of long-distance travel by ship since the time of the Vikings or perhaps even earlier. Their point of origin may have been Europe or Asia.

Some introduced species come from aquarists who dump exotic fishes or algae into nearby coastal waters. If aquarists cannot find a new home for the inhabitants of their saltwater aquarium when they move, or have over-large or aggressive fish such as triggerfish that are wreaking havoc on everything else in the tank, they may think it is kind to let it go free. Such animals and algae may survive and even reproduce in waters thousands of miles away from their native habitat in a region where they have no natural predators. The venomous lionfish (*Pterois volitans*), native to reefs of the western Pacific and Indian Ocean, has been spreading along the Atlantic coast of the United States into the Caribbean, and in 2011 was seen at the Flower Gardens Banks National Marine Sanctuary. These fish escaped from captivity and are reproducing in the wild. Recent studies in Florida and the Bahamas show that lionfish are eating smaller reef fish and severely reducing their populations, and that lionfish are rarely eaten by larger native fishes.

The lionfish, an invader from the Indo-Pacific Oceans, has venomous spines. This fish has lately moved to platforms and shipwrecks in the Gulf of Mexico.

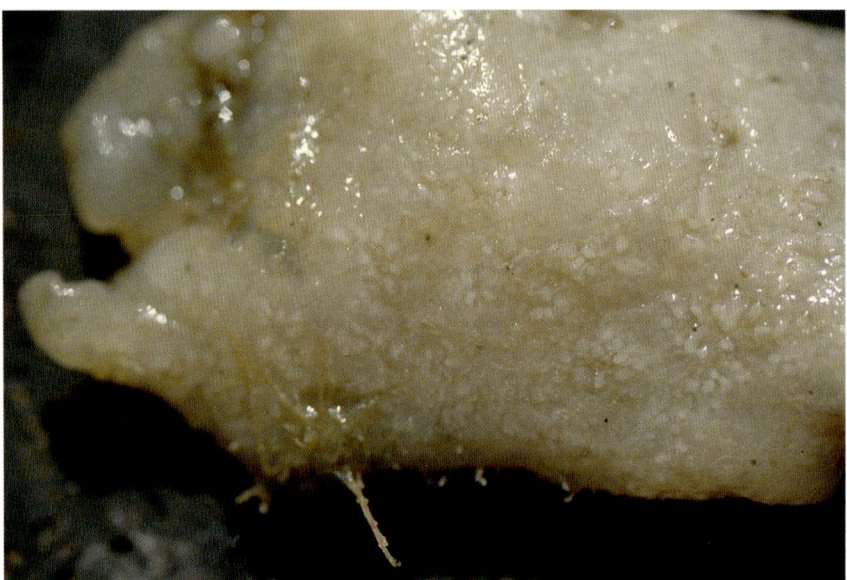

The tiny white dots are the individuals in this rubbery ascidian colony. The individuals form clusters around a central outgoing pore. All the individuals are embedded in a common covering, the tunic. Ascidians can reproduce very quickly on platforms.

Platforms may facilitate the spread of invasive species from platforms to natural reefs. Platforms offer a hard surface on a seafloor composed of mostly mud and silt and could serve as stepping-stones for invasive species. The natural predators of invasive species may be absent from the platform. The potential exists for a "pest" to spread rapidly by "hopping" from one platform to the next. Because of this concern, a number of agencies, especially the National Centers for Coastal Ocean Science (part of NOAA), monitor the spread of exotic organisms.

A species might be native to an area but become invasive under changing environmental conditions. The white ascidian *Didemnum* could have a harmful effect on the biota of platforms. This species settles as an individual and forms buds to create a rubbery flat colony. Experts disagree on whether this is a widespread introduced species or a native species that, having found a good place to settle, is becoming invasive. *Didemnum* contains acids in its tissues, which prevent other species from settling on or near it. Areas heavily encrusted with this ascidian provide fewer places in which fishes can hide or find food.

Platforms are man-made structures and create a habitat that may be favorable for settlement of non-native species. Certain exotic species may be invasive on platforms but do not seem to have an adverse effect on other organisms living there. It is thought that the common striped barnacle, *Balanus reticulatus*, may be an exotic species that originated somewhere other than the Gulf of Mexico. There is general agreement that the white soft coral *Telesto riisei*, found on platforms and shipwrecks in the Gulf, is actually native to the tropical Pacific, as is the orange cup coral *Tubastrea coccinea*. Concern has been voiced that the orange cup coral and other non-native species may spread to natural reefs in the Gulf, but so far, only a few colonies of this coral have been found.

Two species of corals that are considered invaders from the Pacific. The soft coral *Telesto riisei* (A, photo by Dick Zingula) and the orange cup coral *Tubastrea coccinea* (B) are now widespread on platforms in the Gulf of Mexico.

6

RESIDENT ANIMALS

FROM NEARSHORE areas where waters are often turbid to far offshore where the clear, blue-water is found, animals that take up residence on the platforms differ. Water depth, bottom type, light penetration, and distance from shore are factors that determine the members of the animal community associated with a platform.

Seaweeds and the Basis of the Food Chain

The seaweeds (algae) found on platforms in the Gulf of Mexico are usually limited to a few species because of shading and the limited intertidal zone. Most of these are red algae, which, despite the name "red," can be black, red, brown, blue, or even green. These seaweeds tolerate lower light intensities, especially the blue light that filters down to greater depths. Generally, a few tufts of these small seaweeds will be seen in lighted areas at the edges of platforms. In deeper areas, coralline red algae will form a pink crust over the legs and beams of a platform. For the most part, fishes and invertebrates do not eat these algae but they may provide shelter to very small worms and crustaceans.

In the absence of seaweeds on the platforms, the basic food supply of the animals is plankton— a collective term for microorganisms, minute algae, tiny shrimp-like creatures called copepods, small jellyfishes, and the larvae of crabs, barnacles, and other organisms that drift with the current in the water column. Currents may carry planktonic organisms to a platform, where they may be caught in eddies around the jacket, only to die and settle on the framework, or become trapped among shells or other growth. Many animals feed only on certain sizes of plankton, filtering them from the water column using hairs on their arms, comb-like gill spines, special feeding gills, feathery tentacles, or other capture methods. Corals and a few mollusks feed on plankton by secreting gooey sheets of mucus that trap plankton. The mucus and everything on it is eaten, and later sand grains and other inedible materials are passed back out of the digestive tract of the organism. Hydroids, sea anemones, and a few corals sting whatever small animals come too close to their tentacles. Spadefish and small baitfish in the water column rely on plankton, especially larval organisms, as food. Some worms will eat decomposing plankton after it is coated by bacteria, providing, for a worm at least, a protein-rich meal.

Life on Nearshore Platforms

Platforms within about 25 miles of the coast or reaching no more than about 90 feet (27 meters) deep have been called nearshore or coastal platforms. These are most likely to be affected by freshwater runoff from rivers, seasonal cold spells, or easily disturbed silt on or near the seafloor. During summer, the water at 60 feet (18 meters) or deeper can turn silty due to the effects of the nepheloid layer; in winter, the entire water column may be turbid after storms carry silt in from land and stir up the mud of the sea floor. Light barely penetrates the murky water. Near the seafloor, changing silt concentrations can either expose or smother the legs of the platform, causing the cover of attached organisms, none of them more than a few years old, to change constantly. Seasonal distribution of attached animals and fishes varies. During summer, tropical fishes such as cardinalfishes, cocoa damselfishes, and even spotfin butterflyfishes may inhabit the nearshore platforms but in winter these fishes are gone, having died off or moved offshore. Distinct changes in the populations of both algae and sponges can occur between summer and winter.

A visitor rarely encounters much life right at the waterline of a platform. This hostile area can be exposed to air and rain, swept by strong

Dictyota sp. is one of the most colorful algae on platforms. Photo by Dick Zingula.

currents, or smashed by waves. A few oysters, barnacles, and tufts of algae cling to the legs. Two agile and fast-moving crabs pick at algae and debris in this zone. The mottled shore crab (*Pachygrapsus transversus*) has a dark green to reddish, somewhat trapezoidal carapace with distinct grooves running across its width. This is a small crab, with a leg span less than your little finger. The tidal spray crab (*Plagusia depressa*) is larger, with a leg span that can be as long as a person's hand. Its carapace is more rounded and its eyestalks are set in notches. On the front surface of the walking legs of the tidal spray crab are sharp teeth, which are absent in the shore crab. Both of these crabs have keen vision and can scurry quickly. Like skilled mountain climbers, they grip any protrusions with their flattened legs and spiny dactyls.

Nearshore platforms often have species in common with the oyster reefs of nearby bays or the rock jetties along shore. A diver must look closely to distinguish between these animals, which look at first glance like silt-covered tufts and knobs. Barnacles (*Balanus improvisus, B. trigonus, B. reticulatus* and others) commonly form a raised encrusting "rind" on the platforms. Wing oysters (*Pteria colymbus*), tree oysters (*Isognomon bicolor*), and other oysters may be common. The wing and tree oysters secure themselves to the platform by secreting strong attachment threads; other oysters cement one shell directly to the metal. These common animals glean floating particles and tiny animals from the material drifting in the water around them.

Despite its name, the oyster drill (*Stramonita haemastoma*) is just as likely to eat barnacles as oysters or other mollusks. It locates its prey by chemical cues, and then uses a set of tiny teeth to drill a hole through the shell so it can suck out its prey. Oyster drills mate in groups and lay their eggs in tough capsules the size of a grain of corn.

Often, there are dense tufts of colonial animals. These tufts can be mistaken for seaweeds but they lack the red pigment and usually have a texture like plastic or rubber instead of that of a leaf. Hydroids are small animals distantly related to sea anemones. A hydroid is attached to the platform by a stalk, often covered by a tough coating. The largest is *Tubularia crocea*, recognized by its crown of pink tentacles. Hydroids can sting tiny prey. Bryozoans are colonies of extremely tiny animals united into a branched or flat colony. The branching bryozoan (*Bugula neritina*) and other encrusting species can be particularly common in summer to fall. These animals use tentacles to capture tiny particles from the water. Numerous tiny worms and crustaceans live in the "turf" formed by the hydroids and bryozoans.

A small brown or white coral (*Astrangia poculata*) may encrust the platform. Unlike the larger corals of the Flower Gardens Banks, this coral can survive without zooxanthellae in its tissue. It tends to be small and lumpy, not branched, and rarely grows more than the length of a finger.

Sea whips (*Leptogorgia virgulata*), up to 2 feet tall, may be attached near the base of the platform. These soft corals are colonies of sea anemone-like individuals that bud off an original settler that arrived in the plankton. Sea whips also subsist on tiny plankton or particles in the water. Sometimes a barnacle (*Conopea galeata*) settles within an injury of the coral and eventually is nearly covered by the outer tissue of the coral. A small snail (*Simnialena uniplicata*) feeds on nothing but the living tissue of this soft coral. This snail is hard to find because it has the same color as its coral host. Flamingo tongue snails (*Cyphoma* spp.) are also specialists that feed on sea whips and related species.

A few predatory invertebrates inhabit these platforms. Tiptoeing among the hydroids is the sea spider (*Anoplodactylus lentus*), a slow-moving

Two crabs scurry about at the water's edge on platforms. The mottled shore crab, *Pachygrapsus transversus* (A, photo by Angela Witmer), also lives on jetties in Texas, but the tidal spray crab, *Plagusia depressa* (B, photo by Cindy Abgarian), is more common farther to the south in Florida and Mexico.

▶ The wing oyster, *Pteria colymbus*, can occur in large clusters on platforms. Photo by Archie Ammons.

◀ A tree oyster, *Isognomon bicolor*, with small barnacles attached, is one of many common settlers on nearshore platforms. Photo by T. J. Boyle.

▲ Oyster drills (*Stramonita haemastoma*) feed on barnacles and small mollusks on platforms.

◄ *Tubularia* is the largest hydroid found on nearshore platforms. The pink tentacles stick out of the tube.

Bushy bryozoans like this *Bugula neritina* look like algae but are, in fact, colonies of tiny animals. Dense growths of bryozoans are common on nearshore platforms.

Despite the silt, a nearshore platform is densely coated with hydroids: the fuzzy branched organisms seen here. Photo by Dick Zingula.

creature that feeds on the hydroids. Sea spiders have a slender segmented body, a proboscis, feeding appendages and legs with a span of about that of a quarter. This slender animal blends in with the stalks of the hydroids. Sea spiders are very unusual in their breeding habits in that, after fertilization, the female attaches the eggs to the male's special front legs. The male carries the young until they hatch.

Three species of crabs may live on the platforms. The spider crab (*Libinia emarginata*) is a generalist feeder, eating algae, barnacles, small worms, and even small fishes if it can catch them. Juveniles attach bits of algae, worm tubes, hydroids, etc., to special hooked projections on their bodies, thus "decorating" themselves and hiding from predatory fishes. Bigger spider crabs,

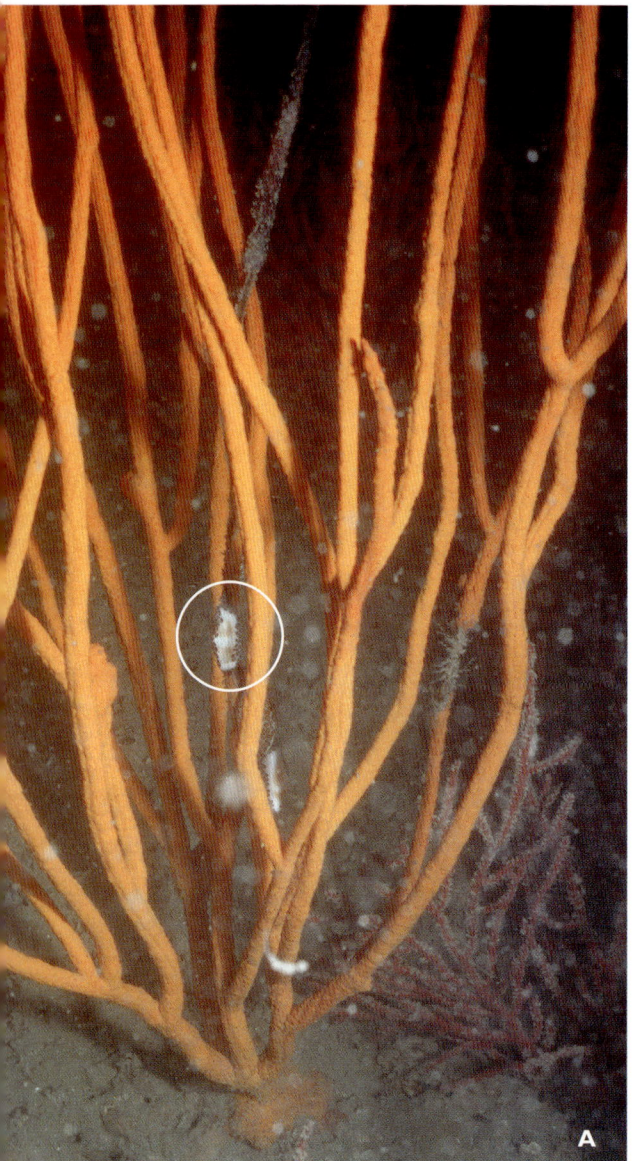

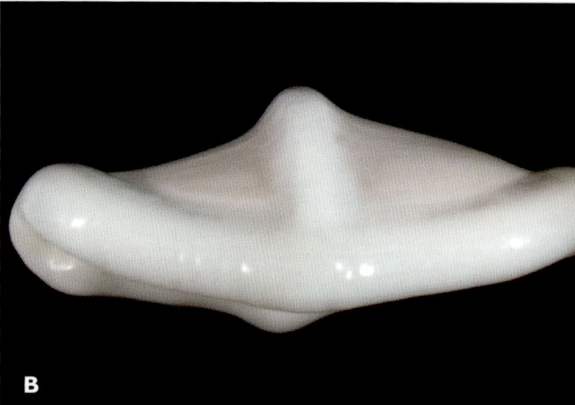

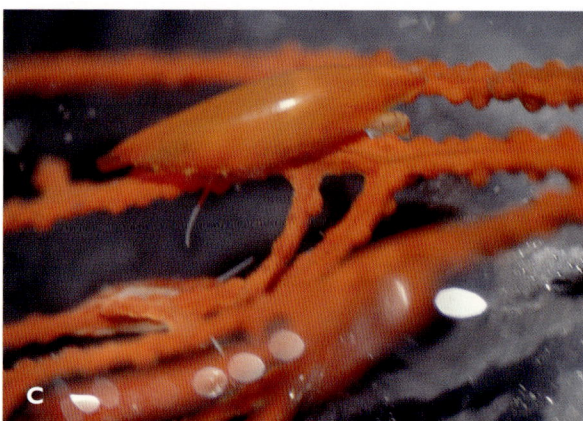

(A) The sea whip *Leptogorgia setacea* is common at the bases of platforms. Note the small white snail, *Cyphoma* sp., feeding on the sea whip at center. (B) Close-up photo of a *Cyphoma* sp. shell. (C) *Simnialena uniplicata* feeds on sea whips and related species. Photo "B" by Fabio Moretzsohn.

with a leg span about the size of a dinner plate, may be encrusted with barnacles. The Gulf stone crab (*Menippe adina,* formerly called *M. mercenaria*) crushes oysters and barnacles, especially smaller individuals or those that have settled away from a large cluster of their species, with its large pincers. The knobbed mud crab (*Hexapanopeus*

paulensis), which, despite its name, rarely lives on mud, is very common among holes and shells. This dime-sized crab feeds on worms, small mollusks, and other tiny prey.

Nearshore platforms may host a large number of fishes, but they are difficult to see in the murky water. Sheepshead (*Archosargus probatocephalus*)

The quarter-sized sea spider, *Anoplodactylus lentus*, climbs among hydroids. Notice the slender pincers, located above the proboscis, and the slender segmented body, characteristic of sea spiders. Photo by Archie Ammons.

forage on small worms, barnacles, crabs, and fishes. Equipped with relatively strong teeth, they can crush through the shells of crabs and barnacles. Belted sand bass (*Serranus subligarius*) tend to stay at the junctions of beams or among hiding places on the bottom. These shy little fish are predators on smaller species, using a wide mouth to "inhale" prey. Gray or mangrove snappers (*Lutjanus griseus*) may congregate near these platforms. They often huddle within the legs of the platform. Adult fish are gray often with a bar running from the tip of the snout to the base of the dorsal fin, but some lack the bar and others are more reddish. Juveniles often live in shallow water among mangroves or near breakwaters. The juveniles may have dark bars on their sides and yellow edges to the fins. Snappers nibble on smaller fishes or invertebrates found on the platforms. Tropical species such as butterflyfishes and damselfishes may show up for a season and then disappear.

Down at the base, various crabs and fishes forage on whatever falls down to the seafloor. The dead barnacle shells may provide shelter for smaller crabs and worms.

Life on Platforms of the Outer Continental Shelf

Past 25 miles from shore, a boat exits the mud line, the dividing line between muddy water and clear, oceanic water. This line can be very sharp, especially near the mouths of larger rivers. Sometimes, a line of drifting seaweed or foam marks this ever-changing division. In a dry summer, platforms closer to shore also may lie in clear water. Distance from the shore is a guideline for what to expect at any given platform.

It is on these platforms that a greater mixture of tropical animals can be seen among species

A spider crab is the largest crab to be found on a nearshore platform. A juvenile spider crab attaches hydroids, algae, and bryozoans to hooked hairs on its body (top). Unless it moves, you cannot see it on the platform. An adult loses the hairs and no longer decorates itself (bottom). Barnacles may settle on these larger crabs.

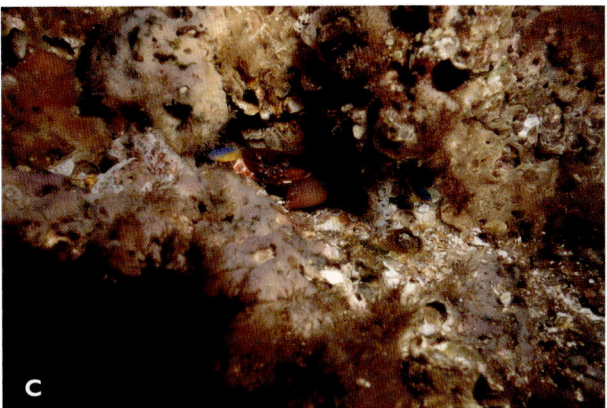

Crabs of nearshore platforms. The knobbed mud crab *Hexapanopeus paulensis* is difficult to see against a bit of shell (A, B). The stone crab (C, D) *Menippe adina* has powerful claws. Notice the shell fragments in front of this crab's hole. These may be remains of the crab's meals. Photo "C" by Dick Zingula.

Fishes of nearshore platforms. (A) A belted sand bass rests among sea whips. The "knots" on the sea whips are the barnacle *Conopea galeata*. (B) Sheepshead occur on nearshore and mid-shelf platforms.

▲ Gray snappers are targets for fishermen on nearshore platforms.

◀ A spotfin butterflyfish inspects the base of a nearshore platform. Notice the silty water. Photo by Dick Zingula.

common on the immediate coast of Texas. Oyster drills and stone crabs coexist with damselfishes, angelfishes, and soapfish. The proportion of tropical species and coastal residents will vary with distance from shore and the other animals that have settled on a particular platform.

It is here that pink barnacles (*Megabalanus antillensis*) are major players in the ecosystem of the platform. Able to attach to vertical surfaces, they can build up massive three-dimensional clusters on the legs and crossbars, often contributing the most biomass and number of individuals on any platform. Whether the ecology of the platform or the problems involved with increasing weight on a platform is of interest, some understanding of the natural history of barnacles is essential.

Pink barnacles vary in size and shape. A barnacle that settles alone on a wide-open space will be shaped like a volcano. Usually, when crowded on the platform, barnacles tend to be shaped like a tower or trumpet: long and narrow at the base and wider at the top. Barnacles will settle atop the dead shells of other barnacles, creating a large cemented mass. Exactly how long barnacles live is unknown, although a lifespan of several years is likely. Their growth rate depends on the local food supply and whether the barnacle grows away from other barnacles or is tightly packed between them.

Adult barnacles do not move much of their bodies other than their feeding arms. These feathery arms beat the water column and collect small particles of food or tiny organisms. Barnacles can adjust their arm beat to local currents and food supply. When disturbed or when they are finished feeding, barnacles seal themselves off by closing a set of movable plates over the opening to the body.

Pink barnacles are one of the animals that seem to do better on platforms than anywhere else in the northern Gulf of Mexico. At extreme low tide, they can be found on rock jetties and breakwaters at Galveston and Port Aransas, a few settle on mooring lines along the coast, and divers have seen them atop the highest ridges of Stetson Bank, 70 miles off Galveston. But why they thrive on the platforms is unknown. Perhaps the abundance of plankton is a factor. Larval pink barnacles may be able to settle more readily on the clean vertical surfaces of the legs than on silty surfaces. Barnacles on the banks and near shore may be scraped off by wave-swept sand, smothered by mud, or dislodged by storms, or be crushed and eaten by predatory crabs. Perhaps algae and sponges that might otherwise overgrow the barnacles are not as prevalent on the platforms as on natural reefs.

Whatever the reason, healthy populations of pink barnacles on platforms support other species as well. Low-growing sponges, hydroids, and red algae may grow on top of barnacle shells. Empty and broken barnacle shells and the spaces between them provide good hiding places for many small crustaceans and worms. Ark shells (*Arca zebra* and others) attach in spaces between barnacles and add to the cluster. The six-armed brittle star (*Ophiactis savignyi*) slithers among the barnacles, collecting fine particles of food with its arms and transferring them to a central mouth equipped with five tiny jaws. The brittle star can regenerate its arms if they are nipped off. Another barnacle inhabitant, the speckled snapping shrimp (*Synalpheus fritzmuelleri*) looks like a tiny lobster. It uses its larger pincer to snap, the sound of which stuns small prey or warns off intruders from its hiding place. Divers are more likely to hear these tiny shrimp than see them. The knobbed mud crab lives among barnacles on these platforms as well as on platforms closer to shore.

Amid very large clusters of barnacles, a diver may find what looks at first to be a barnacle with

The pink barnacle *Megabalanus antillensis* is a characteristic inhabitant of platforms. The shells often are covered by sponges.

jerky movements. On closer inspection, a pair of eyes are seen peering from inside a barnacle shell. This is the tessellated blenny (*Hypsoblennius invemar*). Blennies feed on worms, small crabs, and shrimp on the platform, or small animals carried in with the currents. The tessellated blenny can be distinguished from other blennies by its red spots. This little fish seems to associate almost exclusively with barnacles on platforms in the northern Gulf and on docks in the Caribbean region. Some authors suggest that this species may have been introduced into the Gulf of Mexico with the pink barnacle, but its native range is uncertain. There are records of the tessellated blennies in Florida, the Lesser Antilles, and south to Venezuela or Brazil.

The seaweed blenny (*Parablennius marmoreus*) also lives on platforms but can be abundant on natural reefs. It is about finger-length, and can be dark brown, striped, or even yellow. Both seaweed and tessellated blennies generally stay very close to the platform and "hop" about on their lower fins. Blennies produce eggs that settle on hard surfaces rather than floating free in the water. The young are able to feed soon after hatching, and are attracted to light. They tend to concentrate near the platforms, which are lighted at night.

At the junction points of vertical and horizontal elements of the platforms live larger invertebrates and some bottom-loving fishes. The arrow crab (*Stenorhynchus seticornis*) forages from here out along the legs of the platform, especially at night. This long-legged animal has agile pincers, which it uses to pick morsels of food from cracks. It may look innocent, but it can also snatch and

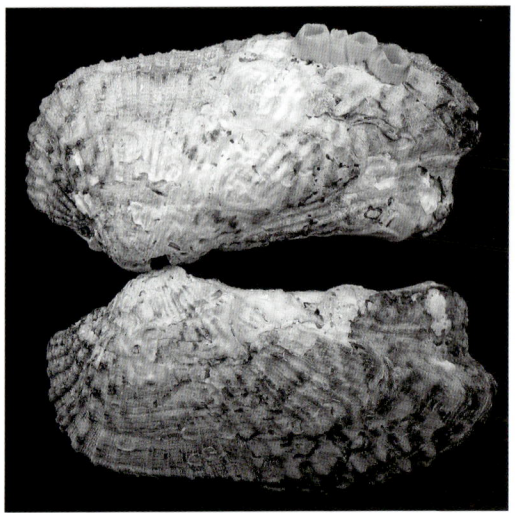

Cleaned of its coating of sponges and other growth, an ark shell, *Arca zebra*, shows its characteristic stripes. These shells are found among junctions of the jacket. Photo by Archie Ammons.

A snapping shrimp, *Synalpheus* sp. Species of this group cannot be identified positively from photographs. Divers hear the "crackling" of their pincers and know that they are on a platform even if they cannot easily be seen. Photo by Linda Pequegnat.

A blenny peeks out of its small hole on a platform. Photo by Dick Zingula.

The tessellated blenny is more likely to be found on a platform than on a reef. Photo by Carol Cox.

eat very small fishes. Stone crabs and oyster drills may live on these platforms, like those closer to shore, and feed on barnacles, ark shells, and other hard-shelled prey. Sea urchins (*Arbacia punctulata*) not only graze on algae but also on small barnacles and sponges. Sea urchins are also found on the deeper parts of jetties from Port Aransas south, but they are sensitive to changes in the salt content of the water and tend to die off if heavy rain dilutes the seawater.

On exposed vertical surfaces, white soft corals can grow six to eight inches out from the platform. These soft corals apparently taste bad so are not eaten, but they provide a protective shelter for smaller animals. This introduced species, thought to have come from the Indian Ocean or western Pacific, is among the most obvious organisms on the platforms. Sponges may also grow here, providing hiding places for brittle stars, small worms, and snapping shrimp.

Many fishes of the platforms stay on or near its immediate surface, and feed on smaller animals there or in the nearby water column. Squirrelfish (*Holocentrus adscensionis*) stay in the shade by day and emerge at night to feed. Rock hind (*Epinephelus adscensionis*), small predatory fish that gulp their prey whole, commonly perch on the beams of the platform. Well camouflaged against the mass of barnacles, sponges, and other organisms, the spotted scorpionfish (*Scorpaena plumieri*) lies in wait. If a smaller fish comes too close, the big mouth opens and sucks the prey inside. This fish has venomous spines in its dorsal fins, so divers should take care and not disturb it.

The cocoa damselfish (*Pomacentrus variabilis*) claims a territory amidst barnacles, sponges, or soft corals and then boldly challenges the intruder. Juveniles have a bright color pattern of blue above with a big spot near the tail and a yellow belly. This color pattern eventually fades to dull yellow and black, and finally to mostly black. The male defends a patch of algae in which females lay their eggs, and then he guards the eggs until they hatch. Males guarding nests may charge even the biggest diver.

Numerous fishes stay near the platform as they feed on the smaller animals attached to it. Blue and French angelfishes (*Holacanthus bermudensis*

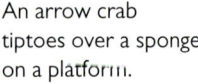

An arrow crab tiptoes over a sponge on a platform.

Echinoderms of mid-shelf platforms. (A) The arm span of a six-armed brittle star barely reaches the diameter of a quarter, but it is one of the most common small animals on platforms. (B) The sea urchin comes in two color phases, black or reddish. Usually it will nestle at the junction of a vertical support and a crossbar. Photo courtesy of Quenton Dokken.

The spotted scorpionfish is the only venomous native fish found on the platforms. Notice the big mouth. This fish swallows smaller fishes or shrimp whole.

may last less than a minute or go on for nearly an hour, depending on the "client" fish species, the degree of irritation, or disturbance from another fish or a diver. As the Spanish hogfish grows, its color changes, its teeth become stronger, and the cleaning habit is lost.

Tiny organisms in the water provide food for sergeant majors (*Abudefduf saxatilis*). These boldly striped fish tend to form large aggregations in the upper thirty feet (10 meters) of a platform. Related to the cocoa damselfish, male sergeant majors also build and tend nests. Their color darkens to warn off intruders and attract females. Anything entering or falling into a nest is removed promptly. Small juveniles eventually drift away among seaweeds or floating debris.

Many fishes are attracted to the abundance of food on the platform itself along with drifting larvae, small crustaceans, and other fishes found in the water among its framework. Large schools of spadefish (*Chaetodipterus faber*) cruise through the legs of the platform, with reports of more than 500 in a single school in the Caribbean. Tagging studies suggest that these fish generally remain near their "own" platform and do not move from one to another. Tiny dime-sized juvenile spadefish may be found among the drifting *Sargassum* and along the jetties on the coast of Texas in spring and summer. These little fish will lie on their sides and look like drifting leaves or trash. The fish move to deeper areas as they grow up.

Large numbers of other fishes also occur around the platforms. Bermuda chubs (*Kyphosus sectatrix*) feed on what they can find, including garbage from visiting boats. If *Sargassum* floats by, they will come to the surface to lunge at the tiny crabs, shrimp, and fishes trying to hide among the seaweed. Creolefish (*Paranthias furcifer*) feed on both plankton and small animals on the platform. They often hover near the platform or rest on its beams.

and *Pomacanthus paru*) feed on sponges. Both of these fishes change their color patterns as they grow. The juveniles bear vivid vertical stripes that fade as they age. French angelfish often travel in pairs, presumably a male and female, although the sexes look alike. Gray triggerfish (*Balistes capriscus*) use their sturdy teeth to crunch through the shells of barnacles and mollusks. The young of these fishes, like those of jacks and other species, may drift with *Sargassum*. Sheepshead also occur on these platforms and nibble on small animals among the sponges and soft corals.

Adult Spanish hogfishes (*Bodianus rufus*) crush prey with their strong teeth. These fish are yellow, dark purple, or a mixture of both colors. The distinctive blue and yellow color pattern of juvenile hogfishes may "advertise" that they are cleaners who will pick debris and parasites off larger fishes, including jacks and barracudas. These larger fishes show their intentions by hovering near the cleaner, tilting the body, opening the mouth, flaring the gills and fins, or otherwise indicating which area needs cleaning. Cleaning

Fishes of mid-shelf platforms. (A) Squirrelfish; (B) Gray trigger (photo by Dick Zingula); (C) Juvenile cocoa damselfish; (D) Blue angelfish.

Fishes of mid-shelf platforms.
(A) Rock hind, a small predatory sea bass. (B) Adult Spanish hogfish. The adult uses strong teeth to graze among barnacles and other shelled prey. Photo "A" by Dick Zingula.

Red snappers (*Lutjanus campechanus*) can form large groups at platforms, with one estimate of more than 26,000 fish at a single platform off the Louisiana coastline. The number of red snappers near a platform depends on its size and complexity, the season of the year, and the water depth. Very small, young snappers (less than one year old) tend to live in areas of low relief

Sergeant majors guard nests on the beams. (A) A nest is made up of fertilized eggs from more than one female, (B) guarded by the father. Photos by Dick Zingula.

on the seafloor—under rock ledges, among junk, or under pipelines. Between the ages of one and two, they move to areas of higher relief. After age three, red snappers often move to natural reefs or deeper waters. Unlike gray snappers, which also may occur on these platforms, red snappers venture out from between the legs of the platform instead of remaining sheltered among them.

Schooling fishes of mid-shelf platforms. (A) Spadefish; (B) Mixed school of Bermuda chubs and snappers (photo by Dick Zingula); (C) Yellow jacks; (D) Brown chromis.

Creolefish pick at plankton in mid-water under a platform.

Larger predatory fishes may remain at a platform or regularly visit it. The most common predatory fish amidst the beams of the platforms is often the barracuda (*Sphyraena barracuda*), which cruises patiently waiting for its opportunity. Barracudas attack other fishes with a quick lunge and a strike with sharp teeth, and may occur by themselves or in loose groups. Members of the jack family regularly pass through the platforms, alone or in groups. Their silvery color, streamlined bodies, and deeply forked tails make them easy to identify. These fast-swimming fishes spawn offshore. The young often have stripes on their bodies, and hide under *Sargassum*, driftwood, or debris offshore from Gulf beaches during summer. As they grow older, jacks move into deeper water, where many form schools. The lucky angler may come across the greater amberjack (*Seriola du-*

merili), which grows to three feet long. These big fish may come to inspect divers. Other species of jacks come by in large schools. These streamlined fish prey on zooplankton and smaller fishes.

Numerous other fishes live on or near the platforms. Studies of the fish communities of platforms off Louisiana suggest that species diversity of reef fishes at mid-shelf platforms is greater than on platforms closer to or farther away from shore. This increased species diversity may be due to the fact that platforms in this area are in close proximity to one another which makes it easier for fishes to move from one platform to another. In addition, where one platform is upstream of another, fish larvae are able to recruit easily to downstream platforms.

Moving down into the deeper areas of a platform, the silty nepheloid layer is encountered. The

Red snappers: fishermen's prizes at platforms. Photo courtesy of Chris Ledford, Texas Parks and Wildlife Department Artificial Reef Program.

The barracuda lurks under a platform.

Greater amberjack, often seen passing through the legs of platforms, are predators on other fishes. Photo by Carol Cox.

level at which this layer is found varies seasonally and even with daily tidal regimes or wave action. At 60 feet deep (roughly 20 meters) or shallower, the water can be clear, with horizontal visibility of at least fifty feet (about 15 meters); at deeper depths you can barely see the end your arm. On deeper platforms, a diver may pass through two or three silty layers between the surface and the seafloor.

In the deeper, silty areas, there are fewer barnacles and more sponges and small corals. The sponges tend to be encrusting types with a low profile. These sponges usually cannot be identified by sight or from a photograph. Two kinds of small corals are usually found: one that is brown and smaller than a dime, the hidden cup coral (*Phyllangia americana*), and one that is branched and usually white, the ivory bush coral (*Oculina diffusa*). These corals do not usually have algae in their tissues. Like their distant relatives, the sea anemones, they catch prey with their tentacles or draw in food particles by secreting sheets of sticky mucus. The sea whip may also be found, attached near the base of the platform.

At the base of the platform is a jumble of old barnacle shells, pieces of dead crabs, sea urchins, corals and shells, bits of metal, lost objects from the platform, and usually lost fishing lures, weights, and line. Visibility is usually poor because silt covers everything. This area is home for small crabs, worms, and other animals that hide amidst the debris. Scavengers forage here. The horse conch (*Triplofusus giganteus*) is the largest gastropod in the northern Gulf of Mexico and it feeds on shelled animals or dead tissue. Its shell may be up to two feet long, but most specimens are much shorter. The giant hermit crab (*Petrochirus diogenes*), a scavenger that inhabits large shells, can provide an attachment place for a sea anemone (*Calliactis tricolor*). Possessing powerful stinging cells for its size (about that of a quarter),

(A) The ivory bush coral *Oculina arbuscula*, inspected by a juvenile blue angelfish, may occur near the base of a platform. (B) The hidden cup coral *Phyllangia americana* is common but hard to see because it is very small.

this sea anemone can protect a crab from attack by an octopus. The spotted porcelain crab (*Porcellana sayana*) lives inside the shell but away from the hermit crab's body. It catches scraps from the crab's meals or particles that the crab kicks up as it crawls. Box crabs (*Calappa* spp.) use their powerful pincers to rip open shells and pick out the meat. An octopus can take advantage of the shells and other debris to make a lair. Various flounders, small sharks and rays, and members of the redfish family scavenge for whatever they can find. Moving away from the immediate vicinity of the platform, the composition of the fauna is that of a typical muddy or sandy area. Commercial shrimp, swimming crabs, flounders and other flatfishes, and tube-building worms are the dominant animal groups.

Life on Blue-water Platforms

The blue-water platforms are located at the very edge of the continental shelf, where the sea floor drops from 250 feet (76 meters) to about 600 feet (185 meters), and finally slopes down into the depths. This area lies 75–100 miles or more offshore. Away from river outflow, these platforms stand in clear water, with light penetrating to hundreds of feet below the surface. Silt gradually settles far below the regions visited by SCUBA divers.

Located far from land, the blue-water platforms also are far from sources of nutrients such as nitrates and phosphates that nourish the tiny algae of the plankton. The supply of plankton in the blue-water is less than that closer to shore. However, even from a boat, some of the most common one-celled organisms can be seen in the water: *Trichodesmium*, algae that look like tiny sparkles.

Below the surface live a few larger but fragile forms of plankton. Various gelatinous forms are common at different times of the year. From spring to fall, the moon jelly (*Aurelia aurita*) and the sea nettle (*Chrysaora quinquecirrha*) float past the blue-water platforms. Both can sting unprotected skin. Currents or storms can carry them into shallow water near the beaches, where they cannot be seen, but near the blue-water platforms, divers can usually see them and hopefully avoid them. Comb jellies (phylum Ctenophora) are fragile stingless animals that use short, sticky

A horse conch forages along the sea floor under a platform.

The giant hermit crab inhabits the largest shells. Hermit crabs and other crabs are scavengers under the platforms.

tentacles to entangle small prey or they gulp their prey whole. Comb jellies may be luminescent—they glow blue if a diver bumps into them at night. Siphonophores are chains of jellyfish-like organisms united into a common colony. Related to the notorious shallow-water stinger, the Portuguese man o'war, these colonies can deliver an irritating sting. Sea wasps (*Carybdea* spp.) have a particularly bad sting. The jelly is the size of a child's block, and usually has only four tentacles.

Although barnacles occur on these platforms, they are not as common as sponges. The reduced supply of plankton in the blue-water may be the reason barnacles do not thrive. Another possibility may be that many sponges are fast-growing, which could smother barnacles, and produce noxious compounds.

Sponges are the most obvious growths that divers see on blue-water platforms. Twenty-seven species of sponges have been reported from a single platform. Sponges generally arrive as microscopic larvae. Most of these are siliceous sponges, equipped with tiny hard parts called spicules that are composed of silicon dioxide, the same material as glass. Many

In the rubbish at the base of a platform, an octopus may have its lair.

As small as a pinhead, a colony of the tiny alga *Trichodesmium* sparkles in mid-water near blue-water platforms. Photo by Angelicque White.

common sponges undergo fragmentation or budding to produce large colonies, but the "individuals," defined by the arrangement of their internal water canals, are indistinct. Sponges feed on small particles that they filter out of the water.

Sponges are extremely variable, ranging in size and shape from thin slimy coatings to bucket-sized colonies. Even experts have difficulty identifying species, because the size and shape of some species differs greatly depending on local water currents or water depth. Many sponges provide habitat for brittle stars, worms, snapping shrimp, amphipods, and even small fishes. Unable to move, sponges do have defenses—either bad-smelling chemical compounds or their sharp, glass-like spicules. Although most fishes and invertebrates will not eat sponges, a few shell-less snails (called nudibranchs) will, as will members of the angelfish family and hawksbill sea turtles.

One of the most common sponges on the platforms, as well as Stetson Bank, is the touch-me-not sponge (*Neofibularia nolitangere*). The second part of the scientific name means "don't touch me." Perhaps the scientist who named this sponge found out the hard way that it stings. This red sponge may look brick red or brown at depth. Its outgoing pores (called oscula) are usually volcano-shaped. If in doubt, any red sponge should not be touched. The tube sponge (*Callyspongia vaginalis*) usually grows on horizontal elements of the platform. Inside the tubes are brittle stars and sometimes small shrimp or arrow crabs may be found.

Divers should watch out, not just for the stinging sponges, but also for other stinging organisms, such as the delicate, fern-like, and often overlooked stinging hydroids. These hydroids often grow larger than their nearshore relatives, with colonies reaching lengths of four inches or more. Sea anemones, as yet unidentified by experts, may be very common on some platforms. The ability of these anemones to sting divers is unknown, but it is best to be cautious and not touch them.

A

Jellyfishes drift past platforms. The sea nettle (A) and the moon jelly (B) are very common species. The sea nettle caught some air bubbles under its bell.

B

▶ The comb jelly, *Beroe ovata*, has no tentacles but instead rows of cilia, called comb rows. Comb jellies are luminescent: they can flash blue light when disturbed at night. This and other very fragile comb jellies drift past the platforms.

▼ The sea wasp or box jellyfish, *Carybdea marsupialis*, has only four tentacles. Divers at blue-water platforms should watch out for this dangerous stinger. Photo by Esat Atikkan.

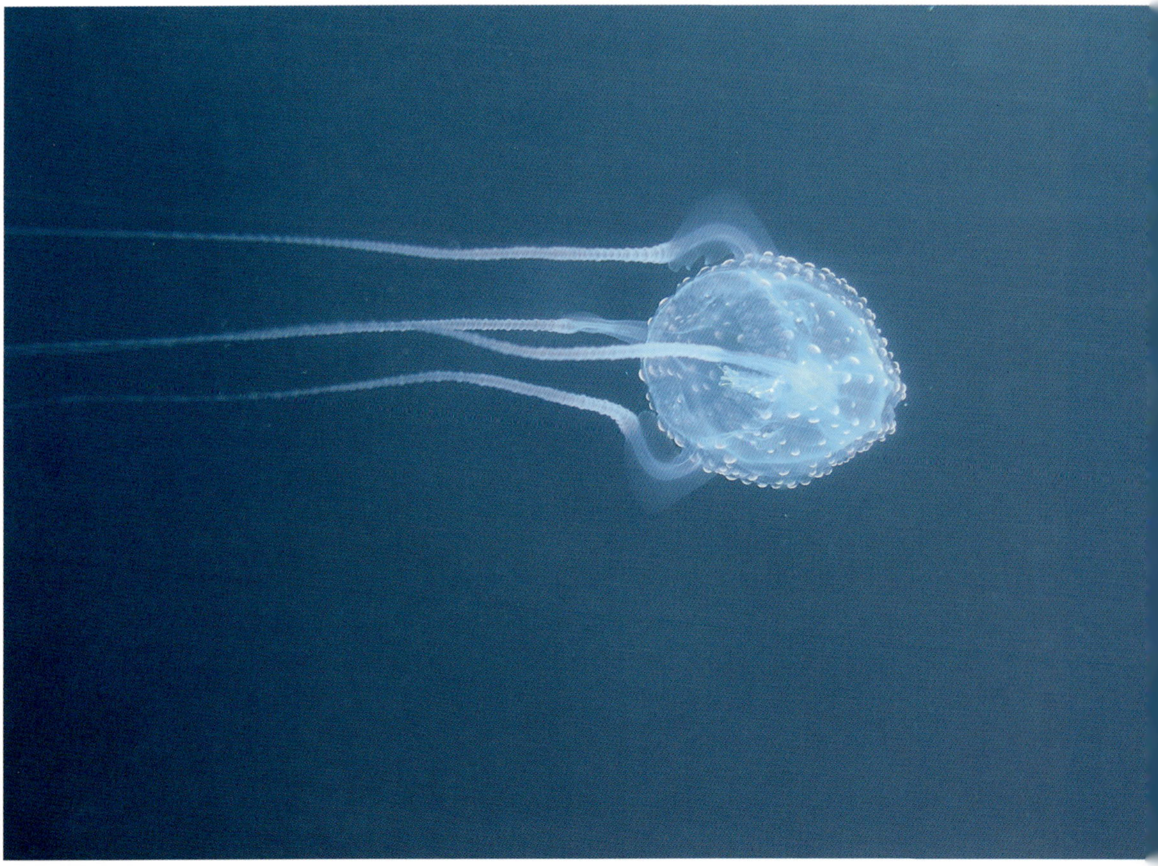

Deeper than 70 feet (21 meters) the feather-like colonies of black coral (*Plumapathes pennacea*) may grow on the structure. In the Caribbean and elsewhere, the hard skeleton of black corals is used to make jewelry. The living tissue forms elongate yellow to golden polyps, which trap tiny food particles out of the water. These slow-growing corals take years to become

large enough for commercial use, and they have been badly overharvested in many areas.

Blue-water platforms have been called "artificial coral reefs." But, only eleven species of stony corals have been reported on platforms compared with thirty-four coral species on the Flower Garden Banks. Of these eleven species, only five species, lettuce coral (*Agaricia agaricites*),

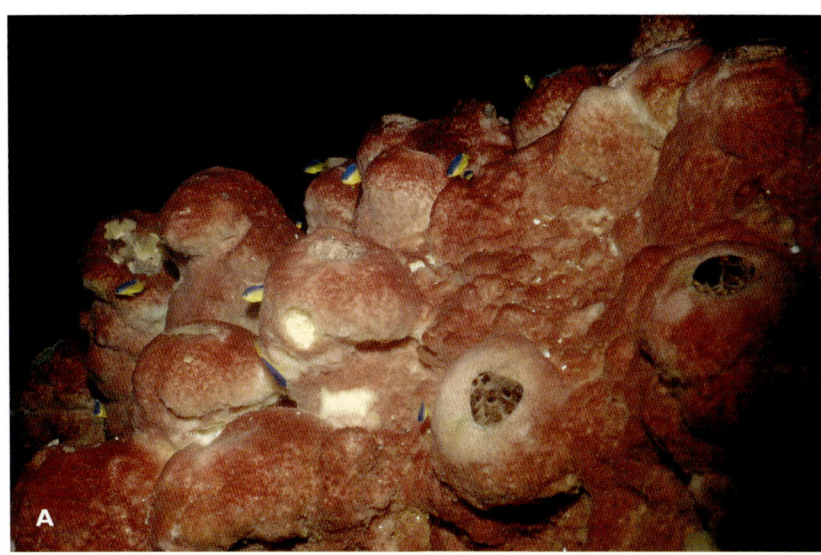

Sponges coat platforms farther from shore. (A) The fire sponge, *Neofibularia nolitangere*. (B) Tube sponges, *Callyspongia* sp. Photo "B" by Dick Zingula.

Stinging hydroid. Although barely one foot high, this colonial animal can pack a nasty sting. This is one of the many animals on a platform that can sting or cut. The diver should keep his or her hands off the platform! Photo by Dick Zingula.

A black coral, *Plumapathes pennacea*, lives attached to platforms at depths of 70 feet (21 meters) or more.

symmetrical brain coral (*Diploria strigosa*), ten-ray star coral (*Madracis decactis*), great star coral (*Montastrea cavernosa*), and orange cup coral (*Tubastrea coccinea*) are commonly reported. Others are known from a few individuals living here and there on scattered platforms. Shading that inhibits their zooxanthellae may account for the absence of some corals and strong currents around platforms may inhibit settlement and growth of more fragile species. There may be competition with barnacles or sponges for space to settle. Some corals will not settle on metal. There is evidence that corals have expanded their populations by colonizing platforms between natural reefs, although genetic studies suggest that the coral populations on the Flower Gardens Banks and those of nearby oil platforms do not intermingle. Between the reefs, the deep muddy seafloor is unsuitable for their growth.

Shell collectors search blue-water platforms for two attractive bivalves (mollusks with two shells, like clams) that have spines that collect sponges and debris, hiding their shells from predators (including humans). Very old shells or those living in a strong current often have spines that have been blunted by years of erosion. When the shell opens, the living animal draws in water to feed and respire, and the living tissue is visible. The jewel box (*Chama macerophylla*) has a pink, white, or yellow shell with short, broad spines. The thorny oyster (*Spondylus americanus*) has a red, white, or pink shell with long spines. In both species, the lower shell is cemented to the platform. Some platforms are encrusted with coon oysters (*Dendrostrea frons*) instead of these two larger mollusks.

As on other platforms, a variety of small worms, snapping shrimp, brittle stars, ark shells, and other small animals live among the sponges. Many of these tiny animals could easily fit on a fingernail, and cannot be reliably identified without a microscope. Among the larger organisms a diver might see are the arrow crab, which is more common on blue-water platforms than closer to shore.

The fluffy, spiraled tentacles of the Christmas tree worm (*Spirobranchus giganteus*) protrude from its shelly tube. The tentacles come in many colors: blue, two-tone black and white, red, and

yellow. These tentacles catch tiny particles from the water that are then eaten by the worm. At the tips of the tentacles are tiny light-sensing structures. If a diver comes too close, the worm retracts quickly into its tube. The rest of the worm, which does not leave the tube, is elongate and segmented.

Divers must beware of the fire worm (*Hermodice carunculata*). It looks like a fuzzy caterpillar and crawls boldly in broad daylight advertising its danger with bright warning coloration and spines. If touched, these spines break off in the skin, causing swelling and irritation. These worms are predatory, grazing on corals and other invertebrates. A fire worm living under a shell or cluster of barnacles can exclude other species that might live there, either by eating them or stinging them

Reef corals rarely settle on platforms. This brain coral (*Diploria* sp.) is on a crossbeam within 30 feet (10 meters) of the surface. Photo by Dick Zingula.

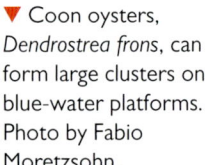

▶ The covering of sponges among the spines hides the jewel box shell from its predators. The arrow shows part of its tissue exposed by the open shell.

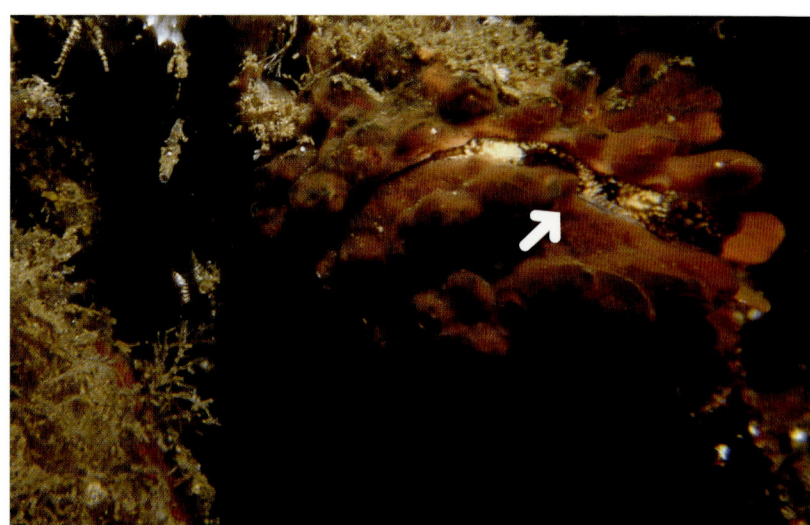

▼ Coon oysters, *Dendrostrea frons*, can form large clusters on blue-water platforms. Photo by Fabio Moretzsohn.

with the spines. Fire worms may have a long life on platforms because their only known predator, a cone snail, does not live there.

In spaces around the collars near the center of the platforms, divers sometimes see juvenile spiny lobsters (*Panulirus argus*). The larvae of spiny lobsters are often found in plankton sam-

ples taken throughout the Gulf of Mexico. The young lobsters eventually settle and hide in holes or cracks or among seagrasses, rocks, and other debris. Active mostly at night, they remain hidden by day. Spiny lobsters often feed on small snails and other mollusks, which they crush with powerful jaws. A platform, devoid of large holes or

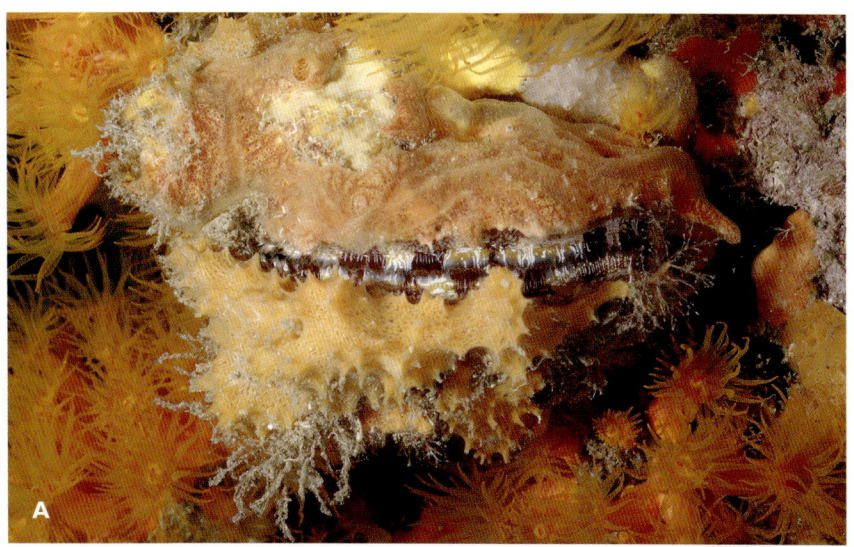

Compare the heavily encrusted thorny oyster above with the one below. (A) This thorny oyster, on a platform, is so heavily encrusted with sponges that all that can be seen is the tissue around the opening of the shells (photo by Richard Sammon). On a reef, the thorny oyster is less encrusted, so the spines can be seen (B). Shells sold as souvenirs have all the sponges removed.

cracks, will be a poor habitat for a spiny lobster. Spiny lobsters that cannot hide fall prey to loggerhead sea turtles and large fish.

Tunicates are small, filter-feeding creatures that may occur singly or joined into colonies. Unlike the invasive white ascidean, the tunicates that live on blue-water platforms seem to be unpalatable, but are otherwise harmless to marine life.

Many of the same fishes that live on other platforms will be found out on the blue-water platforms. Species diversity can be high. According to fish surveys by the environmental organization REEF (Reef Environmental Education Foundation), 117 species of fishes, including fishes that are transients or nonresidents, have been reported at Platform HI A389-A (the platform closest to East Flower Garden Bank). Angelfishes are particularly noticeable among the sponges. If there are few barnacles, the tessellated blenny will not be found on a platform. Numbers of sheepshead and belted sand bass diminish with distance offshore. Amberjack and other members of the jack family may be very abundant.

What lives on platforms in water depths deeper than divers can dive? Abundances of orange coral seem to decline at depths greater than 90 feet (27 meters). Sea whips or sea fans (order Gorgonacea) may be found, along with smaller sponges, small corals, and tubeworms. Deepwater fishes may take shelter among the lower beams. The red crab (*Chaceon quinquidens*), octopuses, sea cucumbers, and fishes typical of the surrounding muddy areas live near the bases of deepwater platforms in the Gulf of Mexico. Large squids and medusae may drift by. Photographs of animals near platforms in the Gulf of Mexico at depths of up to 9000 feet (2750 meters) captured by cameras on remotely operated vehicles (ROVs) are available at the website for the SERPENT (Scientific and Environmental ROV Partnership using Existing iNdustrial Technology) project.

Large members of the grouper family sometimes live in areas below 100 feet (30 meters) near natural reefs or platforms. Warsaw grouper (*Epinephelus nigritus*) and others may come to feed and spawn. There are tales among fishermen of giant goliath groupers (*Epinephelus itajara*), up

Spiny lobsters may be found on platforms, but they have much less shelter than they would find on a natural reef.

Invertebrates of blue-water platforms. (A) Fireworm;
(B) Christmas tree worm; (C) Unidentified sea
anemone; (D) Tunicates. Photos by Dick Zingula.

to six feet long, cruising around the deeper parts of the platforms. These are, of course, the "ones that got away."

Life on Platforms at Night

Most diving on platforms is done by day. Difficulty in reaching the platforms, the potential for getting lost or swept away by strong currents, and, of course, the need for diving experience by day before diving at night all discourage visitors from diving at night. But those fortunate enough to dive at night may stir up luminescent flecks in the water, just like on natural reefs. These are usually one-celled organisms called dinoflagellates, which emit blue light when they are disturbed. They are present by day, too, but in the sunlight their tiny pinpoints of light cannot be seen. At night, creolefishes, damselfishes, butterflyfishes, and others cluster against the platform and nestle in tight spaces among the beams, sponges, and barnacle clusters. Arrow crabs and other crustaceans are more likely to emerge and forage at night than during the day. Brittle stars emerge from their holes and extend their arms to feed. Squirrelfish emerge to forage, sometimes changing color from their daytime red to a pale pink at night. Spotfin butterflyfish (*Chaetodon ocellatus*) change colors dramatically. By day, there is a black line through their eye and a black dot on the posterior edge of their dorsal fin. By night, a large black blotch covers a large area on each of their sides. Red snappers tend to move away from structures at night, presumably to feed. Squid often scoot around the platform, where they feed on small crustaceans in the water.

Lights on the platform or a diver's light may attract swarms of small swimming worms, shrimp-like creatures, and fish larvae. These small animals would normally be attracted to the surface by moonlight. Some of these swarmers will fall prey to the extended arms of brittle stars.

At night, a brittle starfish extends its spiny arms from a hole among sponges, and catches food particles among the sponges.

Many fishes change color at night. During the day (A), the spotfin butterflyfish lacks the brown blotch that it sports at night (B).

7

▼

LARGE VISITING
ANIMALS

LOCATED WELL out to sea, blue-water platforms often attract big fishes, both residents and visitors. Most of the sharks a diver sees belong to the requiem shark family, the Carcharhinidae. These sharks are difficult to identify by sight because their identification is made by comparing the relative lengths and locations of the fins to each other, the length of the snout, and the shape of the teeth. Many sharks are either wary of divers or move so quickly that a diver has little time to see the fine details. Silky sharks (*Carcharhinus falciformis*) can be abundant in spring around HI A389-A, the platform near the East Flower Garden Bank, and other blue-water platforms. Although silky sharks have been reported to reach lengths of 10 feet, most of those seen by divers are much smaller. These fish-eaters usually are not considered dangerous to divers, but they may be attracted to an area where someone has been spearfishing. Other sharks, even the giant whale shark, may pass by the platforms. Divers who see these species near HI A389-A or take photos of them should check the website of the Flower Garden Banks National Marine Sanctuary to find out about ongoing surveys and photographic catalogs of identified individuals.

The manta ray (*Manta birostris*) is a plankton-feeding fish that may be attracted to the lights of a platform at night. Mantas may perform "barrel rolls" as they gulp down the rich "soup" of smaller creatures attracted to lights near the surface. Although they are most likely to pass by blue-water platforms, the occasional manta will stray as far inshore as Port Aransas or South Padre Island if the supply of plankton is especially dense.

Sea turtles may stop to rest among the horizontal beams of a blue-water platform. The most commonly seen turtle is the loggerhead (*Caretta caretta*). This big brown turtle (3 feet long or more) has a large head and a "shell" (the carapace) that is vaulted over the neck instead of resting closely against it. Loggerheads feed on crabs, barnacles, and a wide assortment of other organisms on and above the seafloor. These big sea turtles travel over great distances. They mate off beaches in

72

No one can be sure what one might see during a dive. Even a whale shark may visit a platform and then disappear off into the blue-water.

Sharks may congregate at platforms.

The whale shark opens its mouth wide and gulps down great masses of tiny shrimps and other small animals of the plankton. It has complex filtering gills within its huge mouth. Photo by Simon Pierce.

Largest of the rays in the Gulf of Mexico, mantas are harmless to people. The frontal "horns" can be curled. Recent studies show that individual mantas can be identified by the pattern of spots or other markings on the lower surface of the body.

Florida, then the females go ashore, dig nests, lay their eggs in the sand, and head back out to sea after covering the eggs. The little turtles hatch and scramble back to the sea. Little turtles suffer predation from crabs, fishes, and birds, but those that survive to adulthood can live for decades.

Rarely, the hawksbill turtle (*Eretmochelys imbricata*) may pay a visit to feed on sponges. This turtle has a sharp beak and its shell often has a jagged edge. Legally, the loggerhead sea turtle is designated a "threatened" species whereas the hawksbill, with a smaller population, is an "endangered" species. Both are legally protected from fishing or collecting whole or for body parts, whether the turtles are alive or dead. Grabbing a sea turtle or trying to ride it underwater is harassment, and is prohibited by law under the Endangered Species Act of 1973.

Marine mammals may be seen passing by close to platforms or from boats traveling to or

Loggerheads are the most common sea turtles seen on or near platforms.

from platforms, but are not known to venture underneath platforms or to feed among their beams. Bottle-nosed dolphins (*Turciops truncata*) are found from inside coastal bays to near platforms located on the outer shelf. Spotted dolphins (*Stenella frontalis*) usually live farther offshore. Even sperm whales, recognizable by their size and slanted spout, may come close to platforms. How their activities are affected (if at all) by oil and gas production remains unknown.

Hawksbill turtles feed on sponges on platforms but are less common than loggerheads.

Playful bottle-nosed dolphins often accompany ships. These dolphins may be seen close to nearshore platforms but usually do not venture directly beneath them.

A

(A) Spotted dolphins may pay a visit to a boat near the platforms. (B) A sperm whale, recognizable by its slanting spout and low dorsal fin, is more likely to be seen from a platform off Louisiana than off Texas. Photo "B" by W. F. Samaras.

RECREATIONAL
USE

PLATFORMS provide hard substrate in an area where that kind of habitat is in short supply. The communities of invertebrates and fish that are associated with the structure provide a focus for recreation—primarily fishing and diving—that would otherwise not exist.

Fishing

Structures with vertical relief attract fish, a fact known to every fishing boat captain and boat-owning angler on the Gulf of Mexico. As much as 50 percent of all offshore recreational sportfishing along the upper Texas coast occurs at platforms. Using global positioning systems (GPS) it is easy today to find a favorite fishing spot. In addition, many anglers have a "fish finder" that uses SONAR to locate schools of fish.

Platforms are estimated to have increased the amount of hard bottom habitat in the northern Gulf of Mexico by about 4 percent. Off the coast of Louisiana, the increase in hard seafloor has been much greater, about 10.4 percent. Even for

fishes that are not reef-dependent, platforms offer additional areas for feeding and shelter.

Has the construction of oil and gas platforms increased the population of fishes in a given area? It is difficult to determine whether the number of fishes at any given platform is increasing over time because the fishes are reproducing, surviving, and staying there; or whether adult or juvenile fishes move between natural reefs or other platforms and then congregate at a platform: the population actually is the same, but it is located at only one place instead of several. There is evidence that either or both situations may hold true for different species of fishes. Small reef fishes, such as damselfishes and blennies, stay on a given platform and reproduce there. Snappers and some groupers tend to stay at a platform for a short time, maybe a month or two, and then move away to other platforms, natural reefs, or other shelter, such as pipelines along the seafloor. Juvenile Bermuda chubs and sergeant majors may be carried to platforms in the seaweed *Sargassum* and then gather in the shelter of the jacket and its encrusting life. For migratory species such as jacks, the platforms may concentrate fishes in an area for a short time as they feed before they move on. It is difficult to say where these fish have been feeding during their travels, and thus it is difficult also to say whether platforms, by providing food and shelter, have allowed more of these fish to grow, survive to maturity, and reproduce successfully.

Platforms near shore make the biggest contribution to recreational fishing. The most important target fish are red snappers, of which 4.8 million pounds were taken in 2011. These fish tend to stay near platforms, reefs, shipwrecks, pipelines, and other structures that offer hiding places and some vertical relief. The highest reported catch rates by anglers for red snappers come from oil and gas structures. In the United States today, 90 percent of the commercial red snapper catch comes from Louisiana and Texas, much of it from platforms.

Easy-to-reach platforms off Galveston, Port Aransas, and South Padre Island have received heavy fishing pressure for red snapper. Catches in the western Gulf have been drastically reduced since the late 1980s. Commercial snapper fishermen, operators of shrimp trawlers, and recreational anglers have blamed each other for the decline, but more than fishing pressure is involved. Studies at platforms off Louisiana suggest that the numbers of snappers three years old or older have declined. This decline may be due to fishing, natural mortality, or movement of snappers away from platforms to natural habitats as they age. Studies off Florida suggest that red snappers move away from platforms or other artificial structures at a rate of about 25 percent of the total population at any given structure per year. Yearly recruitment of very small snappers from the plankton is dependent on ocean currents, storms, and abundance of predators, which vary greatly in time and space.

Small red snappers can find shelter in low-lying natural reefs, under pipelines, or even among junk on the seafloor. These little fish are also sometimes caught as bycatch, the incidental catch of non-target species in shrimp trawls. Even if released immediately, small fish caught in trawls may suffer injuries or may be swallowed immediately by opportunistic predators. Any fish caught at a depth of more than 150 feet (46 meters) and brought to the surface quickly may suffer damage to its swim bladder and may die even if released immediately.

Red snappers, caught in the recreational or commercial fishing or as bycatch by shrimp trawlers are the primary sources of red snapper mortality in the eastern Gulf of Mexico, and are also likely to be the major sources of mortality in the western Gulf. Fishing for the big snapper that live in deeper water (over 150 feet) can be particularly damaging. Large snappers can live for more than 40 years. As in many reef fishes, the bigger, old individuals contribute more offspring to the population than do smaller individuals. If big fish are targeted for capture, the result may be fewer new recruits to the population the next year.

An ambitious plan to rebuild the population of red snapper has been enacted through the Gulf of Mexico Fishery Management Council. A total annual allowable catch quota has been set. Shrimp trawlers have been required to install bycatch reduction devices in their nets to reduce incidental capture of fishes. Size limits, including length or weight, as well as limits on the number of fish that can be caught and kept have been implemented, along with seasonal closures. Recent (2009) improvements in fish captures on reefs and artificial structures may indicate that the condition of the red snapper stock is improving. The Council is empowered to evaluate each year's harvest and bycatch rates and make adjustments to meet plan objectives.

Farther offshore is the blue-water, where the big fish are to be found. Trolling for king mackerel, cobia, bluefish, blue runners, and crevalle is popular. Besides these fishes which frequently visit platforms, sharks, dorado, and sometimes billfish and tarpon can move in to feed at a platform.

Really big groupers lurk among the crossbeams at depth. Some of these fishes grow to trophy sizes, such as a 124-pound wahoo, a 121-pound amberjack, and a 302-pound Warsaw grouper, the Texas state records for their species.

Fishermen try their luck near a platform. Photo by Jesse Cancelmo.

A happy fisherman lands an amberjack. Photo by Daniel Campbell.

Diving

New divers may be disappointed to find out that there are no natural reefs or rocky areas close to the shore of the northern Gulf of Mexico. The few rocks and shipwrecks within 50 miles of shore usually are coated with silt. Close to passes, the visibility on breakwaters can be close to zero. The platforms, on the other hand, project all the way to the surface, away from most of the silt. Horizontal beams at 30 feet (9 meters) allow a diver to spend time at a relatively shallow depth instead of going to 60 feet (18 meters) or deeper, where the natural reefs occur.

Diving on platforms in the northern Gulf of Mexico offers advantages that may not be evident to divers accustomed to diving in the tropics. A diving trip to the platforms can be made in a day, without all the hassles of air travel. A diver can spend an hour or more diving on beams at 30 feet or so, well within no-decompression limits. A diver may decide to go deeper and gradually work upwards, ascending slowly and making safety stops. Unlike some of the popular tropical reefs, it is unlikely that the area under a platform will be crowded with other divers. The spearfisherman can bring his gun, which is not allowed aboard commercial aircraft, and find amberjack and other game fishes. Adventurous divers off Louisiana practice a more extreme form of undersea hunting, spearfishing while free diving (i.e., without SCUBA, by holding their breath). Divers can collect clusters of barnacle shells or other ornamental species without the concern of disturbing endangered species or damaging a protected habitat.

Platforms offer opportunities to spend time watching marine life that may not be present on deeper natural reefs. The photographer may admire the play of light beams between the parts of the jacket as great schools of fishes pass by. Divers on blue-water platforms can witness the big fish, rarely seen in the silty water near shore. On a single crossbeam, the diver can see the busy activities of the blennies, brittle stars, crabs, and other small animals amidst the barnacles and sponges.

Divers may travel by private boats from any of the ports along the coast of the northern Gulf of Mexico. Dive shops in Louisiana may offer charters to visit the platforms within 25 miles of the coast, west of the mouth of the Mississippi River. Anchored in deep water near the undersea Mississippi Canyon, these platforms are known for their coating of orange cup corals and the big blue-water fishes that visit the structures. Commercial dive boats usually leave from four ports in Texas: Galveston, Freeport, Port Aransas, or South Padre Island. Divers from Galveston must go out more than 25 miles to find water clear enough for diving most of the time. Platforms about 50 miles offshore often are coated with barnacles and are home to numerous fishes. Boats from ports in Louisiana, Galveston, and Freeport are closest to the Flower Gardens Banks National Marine Sanctuary, about 120 miles offshore of Galveston. A trip to the sanctuary can include a visit to platform HI A389-A, noted for its dense and colorful coating of sponges, tropical fishes, and sharks. Divers embarking from Port Aransas usually head out to platforms 22–25 nautical miles offshore, where the water is clear at depths shallower than 60 feet. Even within a few miles of Port Aransas, a diver can see butterflyfishes and other tropical species, especially during summer. Some platforms in the area have an unusually diverse fauna because they are located near natural reefs, too deep for recreational diving. Off South Padre Island, the seafloor is deeper much closer to shore than it is to the north where the continental shelf is broader, so the boat trip out to a platform with clear water is likely to be much shorter than one out of Galveston. The

"Little Adam" platform and others may host great schools of spadefish as well as a dense coating of barnacles and sponges. As a general rule, the farther offshore or to the south the platform is located, the better the visibility is likely to be.

Before going on a diving trip to the platforms, divers would be wise to have some experience diving in salt water so they can adjust their weights for proper buoyancy. The online site "Oil Platform Diving Recommendations," although oriented toward areas off California, gives good advice on precautions and standard procedures. If aboard a commercial dive boat, be sure to listen to the pre-dive briefing about safety procedures and recommendations for an enjoyable dive.

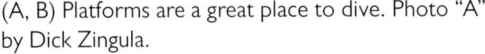

(A, B) Platforms are a great place to dive. Photo "A" by Dick Zingula.

9

THE FATE OF PLATFORMS

PLATFORMS CAN continue to serve as havens for fishes and other marine life even after their oil- or gas-producing days are over. Through the federal "Rigs to Reefs" program and other similar efforts, old platforms can be stripped of anything likely to contaminate the ocean and sunk in sandy or muddy areas. There, the beams and legs can continue to provide habitat for invertebrates and the fishes that eat them or hide among them. A platform may take from fifty to more than one hundred years to finally disintegrate into a pile of rust.

Platforms that are not toppled in place may be moved. Semi-submersible platforms can be towed to shore for cleaning, refitting, and installation at a new site. The upper parts of other types of platforms can be cut off and moved elsewhere, either for reuse in production or as part of an artificial reef. If a platform or some part of a platform is towed in the water, the attached organisms might survive but it is customary to clean a platform before it is moved. If all or part of a platform is moved and installed as an artificial reef,

it must be sunk with careful consideration to the site. The area must not contain active pipelines or be used for military purposes. The platform cannot become a hazard to navigation, although, for a platform to be useful to fishermen or divers as an artificial reef, it must be close enough to a port and shallow enough to be accessible. If it is too expensive or difficult to move all or parts of a platform, a disabled platform by law will be cut off at a depth sufficient for ship traffic, preferably at least 85 feet (26 meters). The wellhead of a platform always is sealed to prevent any leakage.

It is estimated that only about 8 percent of the 100–150 structures removed annually from the Gulf of Mexico have been used in Texas and Louisiana in the "Rigs to Reefs" program. If a whole or partial platform is towed to shore, the life on it will die. Platforms may be cleaned even if they are to be reused elsewhere. Cleaning is a precaution to avoid introduction of invasive pest species from the platform to its new operating area. Exposure to air, silt, and lower salinity water usually kills encrusting animals on the platform.

"Reefing in place" has its own problems. A platform that is left in place with its upper structures intact eventually will rust and may become unstable. Platforms such as these might be severely damaged during a hurricane. Existing navigational beacons must be maintained. There would also be questions about liability in case of a wide variety of accidents, including shipwrecks, oil spills, partial toppling, underwater breakage, or contamination from sources that once were above water.

Platforms may be reefed or cut by use of explosives. Trials off Louisiana suggest that blasting usually does not kill red snapper beyond 150 feet from the blast site, but fish in the immediate vicinity die. As a platform topples, shallow-water organisms are suddenly dumped into the colder, darker water at deeper depths. Sponges and other animals are buried or smothered by silt. A platform toppled on its side loses its vertical profile, and filter-feeding animals may suddenly find their orientation to the currents altered. Unable to move, they will face away from their source of plankton and starve. While schooling fishes such as snappers may move to another platform, reef, or shipwreck, small territorial fishes or barnacle-dwellers such as damselfishes and blennies usually do not leave their immediate areas and will perish. The disruption caused by removal of the platform is likely to alert predators and scavengers, which will move in to feast on dislodged or exposed animals.

In time, a toppled platform can be recolonized and support marine life. At 150 feet (46 meters) or more, deepwater reef fishes, soft corals, feather stars, and other fauna typical of the deeper parts of reefs may move in, settling amidst the legs and beams. Deepwater corals (*Lophelia pertusa*) have settled on decommissioned oil platforms in the North Sea. These corals which provide a habitat for associated crabs and brittle stars, which also occur in the Gulf of Mexico. Because deep-sea reefs are relatively rare, even a small amount of vertical structure may result in increased densities of fishes.

Alternate uses of existing oil and gas platforms have been suggested. Platforms could offer a good base for aquaculture—the constant flow of clean water and elevation above silt would be very good locations to grow edible algae, oysters, scallops, or other organisms. The shallow areas of a platform might be used to grow corals that could be used in restoration of natural reefs. The cost of this kind of reuse is a problem however, because of the amount of money it takes to transport animals, harvested materials, and supplies to and from an offshore platform. In addition, aquaculture facilities would need to be manned, or have surveillance systems to deter poachers.

In cooperation with certain companies and platforms, biologists have used platforms as bases of operations for research. A platform is a fine place from which to make repeated dives without the need for going back and forth from the study area to the shore. Migratory wildlife can be documented on a 24-hour basis. But many practical problems must be solved to use a working platform as a base for research, including arranging transportation for personnel and equipment to and from the platform, and making sure that visiting investigators do not interfere with routine operations. There must also be emergency plans for prompt evacuation in case of a hurricane or if accidents occur.

Some researchers have equipped unmanned platforms with remote cameras or monitoring equipment for research or real-time observations for purposes of education. This kind of equipment is costly and must be sufficiently rigged to stand up to the high swell and strong currents that can sweep across the Gulf. Eventually, equipment must be retrieved for service.

The end of the line? When anything falls from a platform or dies and sinks to the bottom, scavengers are waiting. These deep-sea red crabs (*Chaceon quinquedens*) live at depths of more than 1200 feet (364 m) or more and seem to stay near any hard structures, including the bases of blue-water platforms. Red dots are scale of 4 inches. Photo courtesy of NOAA/Okeanos Explorer.

Platforms might also be used for support facilities for offshore operations in the telecommunications industry. Improved navigational systems for support helicopters already are in place on some platforms in the Gulf of Mexico as well as in the North Sea. Platforms could play an important role in tracking aircraft over the Gulf of Mexico. Air traffic controllers at Bush Intercontinental Airport in Houston have relied on radio reports from aircraft to guide traffic in the Gulf of Mexico. Because there is no ground-based radar coverage over the Gulf as exists on land, a blind spot has been created—a potentially hazardous situation for commercial airliners and service helicopters over the Gulf. At one point, the Federal Aviation Administration proposed installing information broadcasting towers on oil platforms throughout the Gulf. A satellite-guided network of weather and communications equipment could be placed on these towers. This system could track flights with GPS (Global Positioning Systems).

Regardless of the uses to which a decommissioned platform may be put, maintenance problems remain. Will the current operator pay for maintenance of the repurposed platform? How will supplies be delivered? What plans should be made for evacuation and general safety of a platform in the path of a major hurricane? Which user is responsible for liability? What will be done when the platform eventually deteriorates? The operators of a non-producing platform and the BOEM may have a limited opportunity to consider various alternatives regarding its fate.

GENERAL RESOURCES

Exhibits

If the visitor cannot go offshore, there still are onshore exhibits that display aspects of oil platforms. The Ocean Star Museum in Galveston is inside an actual platform along the edge of the ship channel. Visitors can view displays of various types of platforms, see the equipment used, and view the accommodations and equipment of the workers. The upper deck gives the visitor an idea of how large a platform really is. At the Texas State Aquarium in Corpus Christi, Sea Center in Lake Jackson, Texas, and the Audubon Aquarium of the Americas in New Orleans, visitors can see fishes that inhabit a platform in a setting of artificial beams and crossbars. The International Petroleum Museum and Exposition in Morgan City, Louisiana, is a good place to find information on offshore oil and gas production, see displays of the equipment, and walk aboard an offshore drilling rig.

Web Sites

Alternate uses of existing oil and gas platforms: http://www.boem.gov/Renewable-Energy-Program/Renewable-Energy-Guide/Alternate-Uses-of-Existing-Oil-and-Gas-Platforms.aspx
Air traffic control: http://www.faa.gov
Bureau of Ocean Energy Management, Regulation and Enforcement, Gulf of Mexico Region: http://www.boem.gov

This is the first place to search for general information.

Flower Gardens Banks National Marine Sanctuary: http://flowergarden.noaa.gov

Check the "Science" and "News and Events" sections for images, videos, and the latest information on tagging mantas and whale sharks.
International Petroleum Museum and Exposition: http://www.rigmuseum.com
Louisiana Department of Natural Resources: http://dnr.louisiana.gov
National Centers for Coastal Ocean Science, Invasive species: http://coastalscience.noaa.gov/research/pollution/invasive/
Ocean Currents:
Ocean Surface Current Analysis–Real time (OSCAR): http://www.oscar.noaa.gov
Loop Current: http://oceancurrents.rsmas.miami.edu/atlantic/loop-current.html
Ocean Star Offshore Energy Center: http://www.oceanstaroec.com
Oil Platform Diving Recommendations (for the California coast, but still very useful): http://www.ehs.ucsb.edu/units/diving/dsp/forms/Dive%20Plans/platformdiving.pdf
Platform Abandonment: http://www.offshore-environment.com/abandonment.html
Platform Removal Observer Program: http://www.galvestonlab.sefsc.noaa.gov/platforms/
Protected Species (sea turtles and marine mammals): http://www.nmfs.noaa.gov/pr/

See the NOAA/NMFS (National Oceanic and Atmospheric Administration/National Marine Fisheries Service) home page (http://www.nmfs.noaa.gov/) for further information on fisheries.

Reef Environmental Education Foundation: http://www.reef.org

Species lists and population information for reef fishes at sites in the greater Caribbean region.

SERPENT project: http://www.serpentproject .com/default.php

Texas General Land Office: http://www.glo .texas.gov

Texas Parks and Wildlife Department: http:// www.tpwd.state.tx.us

For the "Rigs-to-Reefs" program see: http:// www.tpwd.texas.gov/landwater/water/habitats/artificial_reef/overview.phtml

Texas State Aquarium: http://www.texasstateaquarium.org

Texas Pelagic Birds: http://texaspelagics.com/ Working on platforms: http://www.rigworker .com

Describes the jobs and living conditions on platforms.

Publications

Adams, C. L. 1996. Species composition, abundance and depth zonation of sponges (Phylum Porifera) on an outer continental shelf gas production platform, Northwestern Gulf of Mexico. Technical Report TAMU-CC-9601-CCS. Corpus Christi, Texas: Center for Coastal Studies, Texas A&M University-Corpus Christi. 114 p.

Arvin, J. C. 2001. "Anchored in a river of birds." Texas Birds 3: 27–30.

Boland, G. 2005. "Observations of the antipatharian "black coral" *Plumapathes pennacea* (Pallas, 1766) (Cnidaria: Anthozoa), northwestern Gulf of Mexico." Gulf of Mexico Science 23: 127–132.

Bureau of Ocean Energy Management. *Oil and Gas Leasing on the Outer Continental Shelf.* Available at: http://www.boem.gov/ uploadedFiles/BOEM/Oil_and_Gas_Energy_Program/Leasing/5BOEMRE_Leasing101.pdf

Caillouet, C., W. Jackson, G. Gitschlag, E. Wilkens, and G. Faw. 1980. "Review of the environmental assessment of the Buccaneer gas and oil field in the northwestern Gulf of Mexico." In *Proceedings of the Thirty-Third Annual Gulf and Caribbean Fisheries Institute, San Jose, Costa Rica,* 101–124.

Childs, J. 1998. "Avian diversity and habitat use within the Flower Gardens Banks National Marine Sanctuary." Gulf of Mexico Science 16: 208–237.

Dunn, B., and J. Edwards. 1991. *Diving and Snorkeling Guide to Texas.* Houston: Pisces Books. 90 p.

Fotheringham, N., and S. Brunenmeister. 1989. *Beachcomber's Guide to Gulf Coast Marine Life,* 2nd edition. Houston: Gulf Publishing Company. 142 p.

Gallaway, B., J. Cole, and L. Martin. 2008. Platform debris fields associated with the Blue Dolphin (Buccaneer) gas and oil field artificial reef sites offshore Freeport, Texas: extent, composition, and biological utilization. OCS Study MMS 2008–048. New Orleans, Lousiana: US Department of the Interior Minerals Management Service, Gulf of Mexico OCS Region. 113 p.

Gallaway, B., M. Johnston, F. Margraff, R. Howard, L. Martin, G.Lewbel, and G. Boland. 1981. Ecological investigations of petroleum production platforms in the central Gulf of Mexico. Volume II— the artificial reef studies. San Antonio: Southwest Research Institute. 177 p.

Gallaway, B., and G. Lewbel. 1982. The ecology of petroleum platforms in the northwestern Gulf of Mexico: a community profile. Open File Report 82–03. Washington DC: US Department of the Interior Bureau of Land Management. 91 pp.

Gerwick, B. 2007. *Construction of Marine and Offshore Structures*, 3rd edition. Boca Raton, Florida: CRC Press. 813 p.

Gittings, S., G. Dennis, and H. Harry. 1986. Annotated guide to the barnacles of the northern Gulf of Mexico. TAMU–SB–86–402. College Station: Texas A&M University Sea Grant Program. 36 p.

Hoese, H., and R. Moore. 1998. *Fishes of the Gulf of Mexico*. College Station: Texas A&M University Press. 422 p.

Jackson, W., K. Baxter, and C. Caillouet. 1976. "Environmental assessment of the Buccaneer oil and gas field off Galveston, Texas: an overview." In *Offshore Technology Conference, Houston, Texas, May 8–11*, 277–283.

LGL Ecological Research Associates, Inc. and Science Applications International Corporation. 1998. Cumulative ecological significance of oil and gas structures in the Gulf of Mexico: information search, synthesis, and ecological modeling, Phase I, final Report. USGS/BRD/CR—1997–006 and OCS Study BOEM 97–0036. New Orleans, Lousiana: US Department of the Interior, US Geological Survey, and Minerals Management Service, Gulf of Mexico OCS Region. 130 p.

Lewis, J. R., and A. D. Mercer. 1984. *Corrosion and Marine Growth on Offshore Structures*. London: Society of Chemical Industry/Ellis Horwood Ltd. 156 p.

MacArthur, R. H., and E. O. Wilson. 1967. *The Theory of Island Biogeography*. Princeton: Princeton University Press. 203 p.

Macreadie, P., A. Fowler, and D. Booth. 2011. "Rigs-to-reefs: will the deep sea benefit from artificial habitat?" Frontiers in Ecological

Environments 9: 455–461. http://dx/doi .org/10.1890/100112

Middleditch, B., ed. 1981. *Environmental Effects of Offshore Oil Production: The Buccaneer Gas and Oil Field Study*. New York: Plenum Press. 446 p.

Patterson, W., J. Cowan, Jr. G. Fitzhugh, and D. Nieland, eds. 2007. *Red Snapper Ecology and Fisheries in the U.S. Gulf of Mexico*. American Fisheries Society Symposium 60. Bethesda, Maryland: American Fisheries Society. 396 p.

Peterson, R. T. 1980. *A Field Guide to the Birds East of the Rockies*. Boston: Houghton Mifflin Company. 384 p.

Plotkin, P., M. Wicksten, and T. Amos. 1993. "Feeding ecology of the loggerhead sea turtle *Caretta caretta* in the northwestern Gulf of Mexico." Marine Biology 115: 1–15.

Rezak, R., T. Bright, and D. McGrail. 1985. *Reefs and Banks of the Northwestern Gulf of Mexico: Their Geological, Biological, and Physical Dynamics*. New York: John Wiley and Sons. 259 p.

Sammarco, P., A. Atchison, and G. Boland. 2004. "Expansion of coral communities within the northern Gulf of Mexico via offshore oil and gas platforms." Marine Ecology Progress Series 280: 129–143.

Stanley, D., and A. Scarborough, eds. 2003. *Fisheries, Reefs, and Offshore Development: Proceedings of the Gulf of Mexico Fish and Fisheries Meeting at New Orleans, Louisiana, USA, 24–26 October 2000*. Bethesda, Maryland: American Fisheries Society. 238 pp.

Wursig, B., T. Jefferson, and D. Schmidly. 2000. *The Marine Mammals of the Gulf of Mexico*. College Station: Texas A&M University Press. 256 p.

INDEX

*Note: Page numbers in **bold** indicate maps, figures, or photographs.*

Abudefduf saxatilis, See sergeant major
Agaricia agaricites, See lettuce coral
Alaminos Canyon, 2, 3
algae: requirement for light, 16; coralline red, 30
amphipods, 19
Anaea andria, See goatweed butterfly
angelfishes: 42, 48, 49, 60, 68; *See* blue angelfish, French angelfish
Anoplodactylus lentus, See sea spider
anti-fouling paint, 20
Arbacia punctulata, See sea urchin
Arca zebra, See ark shell
Archipelago Effect, 22
Archosargus probatocephalus, See sheepshead
ark shell, 42, **44**, 65
arrow crab: description and feeding, 43, **46**; on blue-water platforms, 65; at night, 70
ascidian, white, **28**
Astrangia poculata, See encrusting coral
Audubon Aquarium of the Americas, 93
Aurelia aurita, See moon jelly

bacteria, 19
baitfish: arrival at platform, 19; feeding on plankton, 31
Balistes capriscus, See gray triggerfish
ballast water: source of introduced species, 26; scamped, 27
banks: fishing, 11, 18; map, 13
Balanus eburneus, See white barnacle
Balanus improvisus, See rind-forming barnacles
Balanus reticulatus, See striped barnacle
Balanus trigonus, See rind-forming barnacles
barnacles: sessile, 20; life cycle, 20–21; on hard surfaces, 24; larvae in plankton, 30; on nearshore platforms, 32; rind-forming, 32; on blue-water platforms, 59
barracuda, 54, **55**
belted sand bass: on nearshore platforms, 38, **40**; on blue-water platforms, 68
Bermuda chub: arrival at platform, 19; feeding, 48; school, **52;** juveniles carried to platforms, 84
Beroe ovata, See comb jellies
bidding process, 3–5
Big Shell, 17
biological invasions, 26–29
birds: migration, 14; pelagic, 14
black coral, 63, **64**
black tern, **14**
blennies: 43–45; at toppled platform, 91; *See* seaweed blenny, tessellated blenny
blue angelfish, 48, **49**
bluefish, 68
blue runner, 68
boring clams, 11
bottle-nosed dolphin, 78, **80–81**
box crabs, 58
Bodianus rufus, See Spanish hogfish
brain coral, 64, **65**
brittle stars: six-armed, 42, **47**; among sponges, 46, 60, 65; feeding at night, **70**
brown boobies, 14
brown chromis, **52, 53**
bryozoans: on settling plate, **25;** branching, 32, **35**
Buccaneer Field, 24
Bugula neritina, See branching bryozoans
Bullwinkle Platform, 3
Bureau of Ocean Energy Management (BOEM), U.S.: 3, 5, 92
Bush Intercontinental Airport, 92
butterflies, 15, **16**
butterflyfishes, 38, 70, *See* spotfin butterflyfish
bycatch, 85

Other titles in the Gulf Coast Books series: